ENERGY EFFICIENCY IMPROVEMENT OF GEOTECHNICAL SYSTEMS

PROCEEDINGS OF THE INTERNATIONAL FORUM ON ENERGY EFFICIENCY, DNIPROPETROVS'K, UKRAINE, OCTOBER 2013

Energy Efficiency Improvement of Geotechnical Systems

Editors

Genadiy Pivnyak
Rector of the State Higher Educational Institution "National Mining University", Ukraine

Oleksandr Beshta
Vice Rector (Science) of the State Higher Educational Institution "National Mining University", Ukraine

Mykhaylo Alekseyev
Dean of Information Technology Department of the State Higher Educational Institution "National Mining University", Ukraine

CRC Press
Taylor & Francis Group
Boca Raton London New York

CRC Press is an imprint of the
Taylor & Francis Group, an **informa** business

A BALKEMA BOOK

CRC Press
Taylor & Francis Group
6000 Broken Sound Parkway NW, Suite 300
Boca Raton, FL 33487-2742

First issued in paperback 2019

© 2013 by Taylor & Francis Group, LLC
CRC Press is an imprint of Taylor & Francis Group, an Informa business

No claim to original U.S. Government works

ISBN-13: 978-1-138-00126-8 (hbk)
ISBN-13: 978-0-367-37921-6 (pbk)

Visit the Taylor & Francis Web site at
http://www.taylorandfrancis.com

and the CRC Press Web site at
http://www.crcpress.com

Table of contents

Energy Efficiency Improvement of Geotechnical Systems – Pivnyak, Beshta & Alekseyev (eds)
© 2013 Taylor & Francis Group, London, ISBN 978-1-138-00126-8

Preface

This book incorporates the results of research aimed at enhancing the efficiency of geotechnical systems. The contents of the book cover innovative technologies and approaches aimed at improvement of the technical and economic parameters of geotechnical systems functioning. The focus is on mathematical modelling of objects and processes, as well as developing of state-of-the-art approaches and their control algorithms. The book comprises schemata of solving practical tasks related to mine ventilation and electrical circuit operation, cutter-loaders and mining electrical vehicles. It also demonstrates possibilities of hybrid technologies and IT-methods applications to the work of geotechnical systems. Implementation of the methods and technologies presented in the book will allow the reduction of energy resources consumption by geotechnical systems, and enhancement of environmental and economic parameters of their operation.

Genadiy Pivnyak
Oleksandr Beshta
Mykhaylo Alekseyev
Dnipropetrovs'k
October 2013

Energy Efficiency Improvement of Geotechnical Systems – Pivnyak, Beshta & Alekseyev (eds)
© 2013 Taylor & Francis Group, London, ISBN 978-1-138-00126-8

Some aspects on the software simulation implementation in thin coal seams mining

G. Pivnyak, R. Dychkovskyi
State Higher Educational Institution "National Mining University", Dnipropetrovs'k, Ukraine

A. Smirnov
Donbass Fuel Energy Company, Donets'k, Ukraine

Yu. Cherednichenko
Ministry of Coal Industry of Ukraine, Kyiv, Ukraine

ABSTRACT: The features of simulation application of the computer systems in a mining production and the methods of their creation and efficiency application are considered. The UML-diagram of general simulation system forming of the mine which contain the basic mining processes of production is offered. This elaboration is opened, that allows connecting the necessary modules of calculation at the change of mine and geological situation in extraction area. It is oriented on the mutual tying up of all technological chain of separate coal extraction enterprise: processes of working face, transporting of the reflected rock mass and other. Such model enables effectively to manage the mining processes.
The special attention is spared the simulation of wallface. This system enables to show out on the display of computer as simulation the real situation or certain model with the proper admittances in relation to authenticity of results and to carry out monitoring and management of mine production processes directly from the mine surface. Examples of such models are made for the separate mine and geological conditions of the mines SC «Pavlogradvugillya» and SC «Lvivvugillya». The economic evaluation of efficiency work of the simulation systems is executed on the mines of the mentioned companies.
KEYWORDS: Simulating, mining, wallface, coal extraction, computer modeling

INTRODUCTION

Usage of computer systems and development of the proper software allows finding qualitatively new solution of complex problems of mining production and registration of documentation of all technical services of mining enterprises. The concentration of mining operations with introduction of mining equipment of a new technological level and introduction in production the mechanized complexes of a new technological level leads to essential change of a stress-strain state of rocks in a zone of influence of mining faces. Dynamic development of mining operations, existence of various geodynamic fields of tension in the anisotropic environment, the variation geological and a structural compound of the massif on a way of advance of a mining face results to necessity of introduction of adequate changes into technology of coal mining and ways of management of mountain pressure. Overall performance of the mechanized

complexes depends from timely predicted adjustment of technological parameters and the technical structure, adequate change of geomechanical situations on extracting area.

The increase in length of the wallface to 300 m and an extraction column to 2500 m considerably improves use of the mechanized complex according to its moto-resource. Thus, the volume and time of the reserves preparation is reduced, conditions of excavations support improve and other operational costs are decrease. Negative consequences of such approach - is the high probability of change of a technological situation during functioning of a separate mining face. The areas and forms of geo-active zones, gas emission and water inflow to extraction, coal replacements with rocks and so on, concern to them change of a network of geological disturbances.

The power saturation of the mechanized complexes, reliability of their work allows to increase wallface length to effective parameters of economic feasibility. The advance of a mining face geomechanical the situation can change. Mining operations develop dynamically in depth of a rockmass, in the same direction technogenic fields of stress accumulate. Existence of various geodynamic fields of tension in the anisotropic geological environment influences the direction of a resultant vector of tension which has different value depending on its orientation in the rockmass. Development of mining operations can correspond differently to the greatest a vector of intensity and it is reflected in stability of extraction. Work of the mechanized complex depends on timely predicted correction of technological parameters according to change of a geomechanical situation.

For receiving final products, development the passports of mining work, accident elimination plans, calculation the ventilation of a mine, other labor-consuming technical documentation is also necessary. In this regard, in structure of mine (mine surveying, geology, technical service, ventilation, etc.) there are a large number of highly skilled engineers who are engaged in routine work on processing of all-mine information and drawing up documentation of technological processes. At the present stage of computer technologies their work can be significantly simplified and qualitatively improved. For this purpose it is necessary to shift the sufficient volume of settlement and processing part to computers.

The carried-out analysis of all technical documentation of the mine allowed making its general classification by types of works: geological documentation; surveying documentation; passports of mining works; passports of preparatory works & auxiliary mining documentation. Use the computer systems and development the software allows qualitatively new solution of complex problems of mining.

The authors recommend developing absolutely new system of introduction of computer technologies in mining. The platform of the software is accepted the principles of simulating modeling, rather flexible and that can easily adapt to a changing situation on separate sites of the mining enterprise and unite them into general system.

In our case, introduction provides a mine reconstruction, as simulating models of all processes reflecting a geological, mining and technological situation of the enterprise. The role of the engineer advances to the forefront at decision-making at a concrete stage of performance of separate technological process.

THE ESSENCE OF SIMULATION MODELING OF MINING PROCESSES

With the introduction of the electronic computer facilities work with databases, which are an integral part of any modeling, was significantly simplified. This new stage in development of equipment the carrying out calculations allowed to increase significantly complexity not only separate physical and program components, but also of all investigation phases from introduction entrance yielded by the ways of a conclusion of the end results. Applications of the corresponding information platform, operating system and the program of calculation isn't a problem for the qualified programmer.

The complexity of reproduced technological processes of any production depends on algorithm of carrying out modeling and understanding of physical essence. It is important to develop sequence of elementary actions, each of which to turn into instructions clear to the computer, and any computing the task can be solved. Special programming languages, allow changing the separate computing operations into the corresponding program code. The author doesn't want to receive end products in the form of the computer program but only to develop algorithm of actions which become a basis for such activity.

The first simulating models were developed for economic-mathematical calculations and financial office-work in the mid-seventies the last century. They became the essential help when accounting, both insignificant firms, and large corporations. Over time, the area of their application was significantly expanded and these models began to be applied widely at the cumulating of productions, especially it concerns visual reproduction of separate technological processes.

Simulating models are used in the course of computer simulation of productions. In their basis supervision over results of calculations are put at the various entrance values which are set by variable data within an admissible error. However, the greatest share of simulating models make probabilities, that is containing stochastic elements, and demand essential decisions from the operator (Gryadushchiy 2008).

Check of adequacy of simulating model is reduced to comparison of results of calculations from natural data. When the model is realized in a look the block of the scheme, the UML chart or the software product, corrections of mistakes and operational development of inaccuracies is carried out at an appropriate level of modeling, and then data in the best convergence of results are found. These are the main stages of preparation of simulating reproduction of reality.

Simulating models used those cases when reproduction processes, the phenomena and qualities too difficult that it was possible to apply analytical solutions of an objective. Even at application of linear programming, in many cases, passes reality generalization that on received by decisions it was possible to draw valid conclusions.

The mining production is such a system. It is represented in the form of one of the most difficult models which can be reproduced only at the corresponding combination of components. Creation of simulating models of all technological chain at production, transportation and coal processing on the basis of computer providing allows to predict effectively mining technological processes and to make necessary changes in a case of emergence of non-standard situations.

For these conditions simulation as the tactical analysis which helps to make the decision on need and possibility of carrying out real experiment. It is very important for mining process where difficult technologies and expensive equipment are involved. Simulating modeling unlike other methods of modeling represents easily variable system of representation and reproduction of results. In essence, it is final experiment, but not in real and in artificially recreated conditions.

Reasoning from known representations about a type of modeling their set can be presented in the form of the scheme which consists of three components: ideal idea (zone of real values), zone of simulation and zone of discrepancy of results (figure 1) (Grinko 1991).

Perfect idea	Zone of variable data	Zone of accidents

Direction increasing the reliability of simulation results ←

Figure 1. Scheme of an aggregate the simulating model

The structure of simulation modeling is divided into three sectors. "The perfect idea" corresponds real value of object nature. This sector is ideal reproduction of process of modeling. When modeling, actually, has no opportunity to reach 100% convergence to results of "perfect idea"; it is possible to approach the received results to real conditions as much as possible only.

The "zone of variable data" is changeable. Its purpose is the object adaptation - model to real object by means of comparison of fictitious results of model to real results in the conditions of mine. It is necessary in the course of modeling.

The "zone of accidents" is sector of modeling at which the validity of process of modeling isn't observed. It is considered that here has no compliance of results of modeling.

The direction of increase of reliability of results of modeling is indicated the need as much as possible precisely to correspond sector by "perfect idea" is real value of modeling.

The authors suggest approaching more widely to simulating modeling and it is essential to interpret zones of simulating modeling (figure 2).

Perfect idea		Zone of variable data		Zone of accidents	
Perfect compliance of results	Area acceptable convergen	Zone of satisfactory convergence	Critical compliance	Zone of unacceptable results

Direction increasing the reliability of simulation ←

Figure 2. Expanded Scheme of an aggregate the simulating model

When modeling the zone of simulation has to reproduce real data of modeled object as much as possible. Results of researches in "a zone of satisfactory convergence" and "to a zone critical compliances" are expedient for receiving preliminary results about process of modeling and are applied only to acquaintance with modeled by object. In that case proportions of the geoterich sizes or physical properties of object-nature and object-model remain only. They are suitable at the solution of the general fundamental tasks. For example such admissibility can be accepted at visual reproduction of object model. In that case we only reproduce these processes with preservation of the geometrical sizes, physical execution, and we keep proportions in reproduction of the sizes of the capital equipment in the wallface. The same situation is at simulation the schemes of ventilation, the plan of elimination of accident, etc. We don't reproduce the large-scale sizes of extraction, and only we set their schematic location. In a zone critical and satisfactory compliance of results we can examine object interesting to us only.

"The zone of admissible convergence" and "ideal compliance of results" reproduce researches which demand accurate compliance of results of modeling to real results of nature. In these sectors are applied only checked mathematical mechanisms and are used a big number of variable checks.

BASIC DATA FOR MODELING

Application of simulating modeling in the conditions of modern mine is based on simulated reproduction of all of its qualitative and quantitative characteristics: type and mine structure; technologies of production, scheme of opening of a mine field; schemes of preparation of mine or parts of mine fields, extract systems; sequence of working off of layers; planned schedule of extraction of reserves; type of underground transport, scheme and way of ventilation; numbers of blocks of a mine field in simultaneous work; numbers of the horizons, floors, panels in a mine field; number of extraction fields, etc. Common approach to a choice of parameters and elements of mines is that at the beginning is planned previously values

and characteristics of parameters of mine, and then these values and characteristics are consistently specified on the basis of calculations and mutual coordination.

Some parameters and schemes can be unambiguously determined by mining conditions. If various values of parameters and various decisions are possible, options which are estimated by the accepted criteria of efficiency. The compare parameters of mine on the basis of the technical and economic analysis and specifies in the course of simulating reproduction. All these processes logically connected in system of the automated control of the mining enterprise (figure 3) that allows to analyze systemically a condition of mine and to introduce necessary amendments into mining process.

Creation of simulating model of all mine in the specified volumes demands not only big efforts at the level of the developed structure of programmers, but also high system opportunities of the software. Therefore at design of such difficult information system it is offered to break it into components, each of which is considered separately. Two various ways of such splitting into subsystems are possible: structural (or functional) splitting and object (component) decomposition.

The essence of functional splitting consists of system "the calculation program - variable basic data - algorithms of performance of actions". At functional decomposition of program system its structure can be described by the flowcharts which knots are "the processing centers" (functions), and communications between knots describe movements of data.

Object system division still call component that found reflection in the special term:" the development based on components". Other principle of decomposition is thus used - the system breaks into "an active essence" - objects or components which interact with each other, communicating and acting to each other in the relation "the client - the server".

The server is the dispatching service of mine or online support of information product by his producers through worldwide network the Internet. The message which the object, defined in its interface can accept. In this sense the message parcel "object - the server" is equivalent a call of the corresponding method of object.

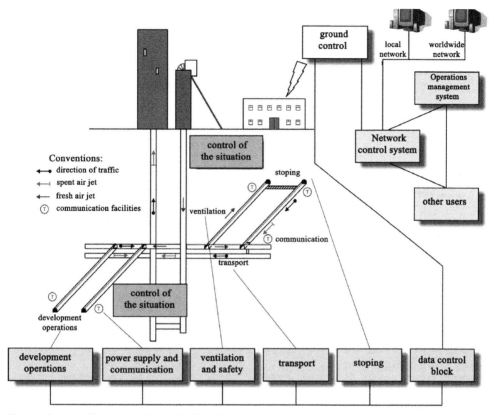

Figure 3. System of the automated control of the mine

Authors suggest accumulating on system "mining works», which is to rise by level of simulation wallface: mining face, preparatory workings, bench connections with drifts and lithologic a structure of lateral rocks.

For this system basic data for modeling are two major factors - a mining-and-geological structure and the technological situation in a wallface.

The lithologic difference is put by means of display of the planes of division on stratification of rocks (figure 4). Difference of points in the vertical plane (U) on length of an extraction column (L) is described by means of "soft" mathematical models in the form of curve various orders:

$$U = f(L).$$

On received on two extraction drifts (onboard and conveyor) the geological structure of the rockmass is formed. It is corrected by means of connection to modeling of these prospecting wells, containing in limits of a considered extraction column.

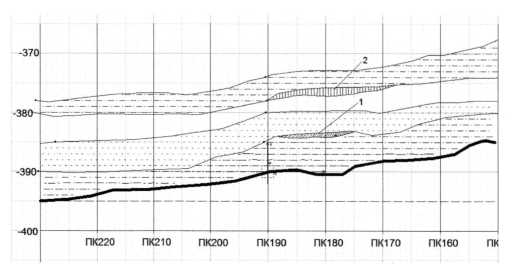

Figure 4. Cross section of geological thickness by one of the site drifts

Geological disturbances, complexity of a structure bring to very difficult the mathematical mechanism of input of entrance data and construction on the equations of a geological difference of the rockmass. Therefore the authors suggest entering data manually. Thus, it is possible to reflect rather in detail sharp changes in a geological structure of the massif. It is promoted also by the software applied to reproduction of graphic objects of interpolation of lines, creation of isohypses etc.

PROCESSES OF THE HIGH-MECHANIZED EXTRACTION

For modeling the process of performance of operations so-called charts of activity are used. The graphic component applied in them is similar to condition charts as on activity charts also there are designations of stationary conditions and high-quality transitions of system. Difference consists in semantics of conditions which are used for representation not static character of system, and dynamics of its change and in absence on transitions of a signature of events. Each action of the chart corresponds to performance of some elementary operation, and transition to the following condition works only at end of a definite purpose. Graphically the chart of activity is submitted in the form of the activity schedule which tops are actions, and arches - high-quality transitions from one condition to another.

Charts of activity play an important role in understanding of modeled processes. Traditional flowcharts of algorithms used for this purpose possess serious restrictions in representation of parallel processes and their synchronization. Application of paths and objects opens additional opportunities for evident representation of processes, allowing specifying simulated processes.

Thus, process of the object-oriented analysis and design of difficult systems is represented as sequence of iterations of descending and ascending development of separate charts. In case of the standard project the majority of details of realization of actions can be known in advance on the basis of the analysis of existing systems or the previous experience of development of systems prototypes. For this situation ascending process of development will be dominating. Use of standard decisions can significantly reduce time of development and prevent possible mistakes at implementation of the project.

When developing the project of the new system which process of functioning is based on new technological decisions, the situation is represented more difficult. Namely, prior to work on the project can be unknown not only details of realization of separate actions, but also their content becomes a development subject. In this case descending process of development from more general schemes to the charts, specifying them will be dominating.

The activity chart as well as other types of initial charts, doesn't contain means of a choice of optimum decisions. When developing difficult projects the problem of a choice of optimum decisions becomes actual. Rational expenditure material and the labor costs spent for development and operation of system, increase of

6

its productivity and reliability often define the end result of all project. In such situation it is possible to recommend use of additional resources and the methods focused on analytics-simulating research of models of system at a development stage of its project. The chart of activity of system of the automated design by mining operations (figure 5) allows unwinding all actions which we will carry out in the course of modeling and their sequence in space and time.

At the first stage of activity passes the analysis and reproduction of mining-and-geological conditions of a bedding of stocks. For this purpose graphic and numerical information serves.

To graphic information belongs:

- three-dimensional display of a mine field (display of contacts of geological layers with a binding to system of coordinates);
  - drawing of geological disturbances;
  - receiving geological cuts.

To numerical information belongs:

- thickness of a coal and of a rock layer, their strength, water cutting, existence of inclusions, etc.;
- numerical characteristics of a bedding of coal layers (capacity, hade, durability, fracturing, etc.).
- angle tectonic and natural fracturing, orientation of rather mining works.

To reproduction of a mining situation at working off of stocks it is used the statistical data connected with process of coal mining. Therefore at this stage the database and the analysis of technology of working off of stocks is applied.

Databases of technical providing include:

- types of fastening of excavations;
- type and structure of the mechanized complexes;
  - ways of protection of excavations;
  - mining equipment.
- ways of fastening of connections of a bench with drifts

In the analysis the technology of working off of stocks has to be in details considered and reproduced the following information:

- schemes of opening of a mine field;
- way of preparation of reserves;
- development system;
- direction of reserves extraction.

At the second stage the analysis of existing equipment and technology of production of stocks of coal on mine is carried out. It is carried out by a well-known technique of design of productions on mine:

- choice of the scheme of opening of a mine field, way of preparation and system of extraction
- mining-and-geological condition.
- three-dimensional model of a mine field
- chosen mining equipment.

At a stage of increase of efficiency of production of stocks of coal the choice of versions of possible technical and technological solutions is made and settled an invoice their economic feasibility.

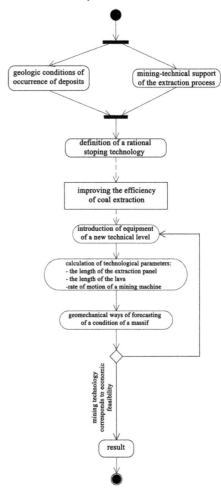

Figure 5. Chart of actions of the system of the automated design

Each technological introduction has to be based on development scientific and technical progresses and the latest development in system "science - design - production". Therefore, at this stage the equipment is carried out most suitable. Main objective of this stage is the analysis of technology of development for elimination of "bottlenecks" and increase in capacities.

Further for each possible variant it is carried out the calculation of technological parameters of production:
- length of an extraction column;
- wallface length;
- speed of of the cuttingloader, etc.

According to chosen technological parameters of working off of stocks zones PGT pay off, are displayed in three-dimensional model and checked on compliance of possibility of application and safety of performance of technological processes.

Process of a choice of the corresponding technical and technological improvement is carried out to full economic feasibility from the point of view of the separate mining enterprise. Iteration of recurrence of conducted researches depends on complexity of the system, an available know-how and necessary accuracy of the received results.

All system of simulating representation of system of modeling of the high-mechanized extraction of thin coal layers consists of the chart of classes which serves for representation of static structure of model of system in terminology of classes of object-oriented programming. The chart of classes, as a rule, is static representation of object model or system model. At detailed level they are used for generation frame a program code in the set programming language, and also for generation of SQL DDL of the offers defining logical structure of relational tables.

For the description of dynamics of system behavior charts which share on charts are used:
- conditions;
- activities;
- interactions (consist of charts of sequence and mutual exclusion charts)
- realizations (consist of component charts and expansion charts).

It can reflect interrelations between essence and evident results, and also describe internal reproduced objects and structure and types of the relations. On this chart information on temporary aspects of functioning of all system isn't specified. The chart of classes is further development of conceptual model. The chart of classes is the certain count which tops are "qualifier" elements which are connected by the various structural relations. The chart of classes can contain interfaces, packages, communications and separate software packages. When speak about this chart, mean static structural model of designed system. However, it depends on changes in space and time.

Process of development of the chart of classes takes the central place in object-oriented modeling of difficult systems. From ability it is correct to choose classes and to establish between them interrelations often depends not only success of process of design, but also productivity of implementation of the program. As the practice shows, each programmer in the work seeks to make use to a certain measure of already saved up personal experience when developing new projects. It is caused by desire to reduce a new task to partially solve to use the checked fragments of a program code and separate components.

Such approach allows reducing significantly terms of implementation of the project. After development of the chart of classes process of modeling can be prolonged in two directions. On the one hand, if behavior of system trivial, it is possible to start development of charts of cooperation and components. For difficult dynamic systems the behavior represents the most important aspect of their functioning. Specification of behavior is carried out consistently when developing charts of a condition, sequence and activity.

Developed chart of classes of system of functional design isn't the final product in the form of a separate software package in language of computer programming. It contains all necessary components for simulating reproduction of mine at high-performance extraction of thin and very thin coal layers. The authors managed to realize only separate parts and to connect them in uniform simulating system.

SIMULATION REPRODUCTION PROCESSES OF MINING

The authors offer completely recreate a computer simulation model of the entire mine or her individual site that enables effectively to monitor mining operations and proactively make the necessary technological changes in the process of coal mining. Transverse sections of the geological strata for roadway and structural columns studied areas is the initial information for the simulation of the geological structure of the rockmass and mines that are within these limits. For example, we give a reflection on the plane simulation of lithologic structure (Figure 6).

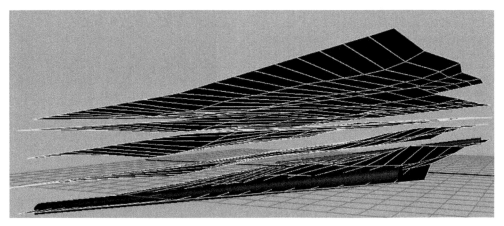

Figure 6. Simulation of a lithological structure

To determine the length of rational walface authors used a technique based on determining factor geological and structural buildings massif (section 4). Part of this work was to mine "Stepova" JSC "Pavlogradugol" using mechanized complex KD - 90. Screenshots of the software are shown in Figure 7.

Figure 7. Screenshots of simulating representation of a technological situation on a mining stope site

The forecast of the condition of the rockmass is carried out by the analysis of geological and surveying documentation and monitoring mining pressure. It is basic data for determination of technological parameters of a mining face, but doesn't consider a condition of mining equipment. Technical characteristics of the mechanized complexes are focused on average values of behavior of the rockmass and consider frequent changes of mining-and-geological conditions insufficiently. The technical system of the mechanized mining face works independently and its condition is partially estimated when performing scheduled or repair work.

Timely replacement of the component of technical or technological system depends on anticipation material service. Its actions have discrete character, coincides with planned indicators of mining. Between separate services of mine not coordinated actions which make a problem when planning intensive development of mining. Absence of algorithm of interaction of separate subsystems (geological and surveying, monitoring of geomechanics, transport and other) which defines a form and geometrical parameters of technological system and allows to interfere and correct quickly work at intensive extracting coal seam.

Reasoning from a condition of coal-mining area of Ukraine and need of its stable further development, development of the scientific principles of formation of simulating models in mining is very actual task. The results of works is planned to introduce at the enterprises of mining branch of Ukraine, the preference is given to coal mines of Ministry of Fuel and Energy of Ukraine.

CONCLUSIONS

The scheme of simulating model allows expanding considerably a scope of computer programs at simulated reproduction of difficult processes high-mechanized production of thin and very thin coal layers. Creation of computer simulating models of all complexes of mine or its separate structural units gives the chance to operate effectively the mining process. Simulating is very effective, flexible and dynamic method of representation of difficult systems that is an integral part at reproduction of difficult processes of mining.

Possibility of a combination of various modules within simulating model which can be already existing information appendices or author's development of certain researchers significantly is expanded by possibilities of this type of modeling.

Creation of simulating model of mine doesn't aim at development of a final information product, in the form of a package of computer programs, and only accurate system of its realization, including all mathematical mechanisms, used in calculations.

REFERENCES

Gryadushchiy, Yu.B., Dychkovskyi, R.O & Ruskych, V.V. 2008. *Software Simulation in Thin Coal Seams Mining.* 21-st World Mining Congress & Expo 2008 Krakow, Katowice, Poland.

Grinko, N. K., Ustinov, M. I. & Osipov, T.V. 1991. *Simulation model of a mine, as the tool for development of the forecast of scientific and technical progress at underground coal mining.*

Energy Efficiency Improvement of Geotechnical Systems – Pivnyak, Beshta & Alekseyev (eds)
© 2013 Taylor & Francis Group, London, ISBN 978-1-138-00126-8

Normalization of voltage quality as the way to ensure energy saving in power supply systems

G. Pivnyak, I. Zhezhelenko & Yu. Papaika
State Higher Educational Institution "National Mining University", Dnipropetrovs'k, Ukraine

ABSTRACT: The article deals with the problem of electric power quality. The problems in power systems accompanying losses in voltage quality are analyzed. The main causes of quality losses are identified. The technique of detection and evaluation in reference to additional loss in electric networks is given. The conditions which provide the emergence of resonances in the networks with higher harmonics are described. Measures to limit the higher harmonics are analyzed. The variety of harmonic filters is described. Findings which prove energy savings observed at normalized quality rating are presented.

1 INTRODUCTION

Coal industry is considered the top priority in the economy of Ukraine, specifically in forming the index of the gross domestic product. Vertically integrated company DFEC (Donetsk Fuel and Energy Company) is the power engineering leader in Donetsk and Pridneprovs'k region. Heavy investments have been made in the modernization of mining and energy equipment to improve the efficiency of coal mining since the corporation of coal mines "Pavlogradugol" has joined DFEC. The electric drive with frequency response has become more widely used at powerful electricity generating plants (main mine fans (MMF), all types of hoisting units, belt haulage).

Intensive installation of power converters in the power-supply system of mines results in negative consequences such as electromagnetic interferences (EMI), mostly harmonic nature. Higher harmonics in power systems were considered to be the main form of electromagnetic interference up to the end of the previous century. At present, due to the wear and tear of equipment and significant deterioration of the main facilities of electric-power industry (by 70-80% in Ukraine and Russia), the voltage quality derating has become an urgent problem. Voltage quality increase in electric power systems is a matter of great importance in advanced economies as well. It is estimated (Zhezhelenko 2004; Zharkin 2010), that EU industry and community on the whole lose about 10 billion euro's per year due to power quality problems, while the costs of preventive measures are less than 5% of this amount.

It should be noted that the problem of power quality in the mains of Ukraine is extremely specific. Thus, for example, in all industrialized countries of Western Europe, the connection of powerful non-linear loads which can distort the current and voltage curves of power supply is allowed only if the quality of supplied power satisfies specification limit and there are appropriate adaptors. At the same time, the total capacity of the newly introduced non-linear loads should not exceed 3 - 5% of the total load capacity of the power company (Zhezhelenko 2012). The situation is quite different in Ukraine where connection of powerful single-ended and non-linear electrical receivers is chaotic. The developed system of higher power rates, imposed for deteriorating the voltage quality by the consumer, is not efficient, which results in systematic voltage derating, increased electric loss, the occurrence of resonance processes at high frequencies in the network capacity.

The paper aims to analyze the current situation as regards power quality in the networks which supply profitable mining enterprises and to identify the conditions of resonance emergence. Special attention is paid to the evaluation of power consumption modes at mines where adjustable filter compensating units are installed.

2 THE PROBLEM OF HIGHER HARMONICS

Power quality rating (PQR) is one of the main problems concerning electromagnetic compatibility of consumers.

Higher harmonics in electrical networks are undesirable due to a number of consequences for the power supply system of the mine, i.e. they have a negative impact on the technical state of electric equipment and deteriorate economic parameters of its operation. This contributes to the incidental loss of power and energy, which in turn affects the thermal regime of electrical equipment, complicates compensation of the reactive power with static capacitor banks, electrical machinery and apparatus service life is reduced due to the accelerated wearing and tearing of insulation, system and networking equipment failures and malfunction of telemechanic equipment is observed.

The impact of higher harmonics is implicit and has a cumulative effect, so the consequences, such as the insulation defect of electrical machines and cables, could be expected within a certain period of time. Voltage waveform distortion affects the occurrence and behavior of ionization processes in insulation. In the presence of gaseous inclusions in insulation the process of ionization occurs, the physical essence of which is to generate a space charge and its subsequent neutralization. The neutralization of the charges correlates with energy dissipation which results in electrical, mechanical and chemical effects on the surrounding dielectric. The above mentioned facts are considered to be the causes for local defects in the insulation which reduce the service life of the insulation. In the presence of higher harmonics in the voltage curve the insulation wear and tear process is more intensive, especially in cables and capacitors. Some authors argue that if the value of the higher harmonics accounts for 5%, then tgδ of capacitors doubles after two years of their operation.

Additional loss both of active power and electric power is detected during the flow of harmonic currents through the network elements and the electrical equipment of consumers.

The greatest value of active power loss caused by the higher harmonics is observed in transformers, motors, generators, overhead power lines and cable transmission. The increase of active resistance of these elements caused by the frequency step is proportional to the magnitude $\sqrt{v}$, although it is somewhat of imprecise approximation (Zhezhelenko 2004). In some cases, additional loss may cause unacceptable overheating and failure of electrical equipment. The value of the additional loss in active power and energy is determined by the mode of operation of electrical equipment and the level of the higher harmonics in the network.

Additional value of active power loss in the air and cable transmission is defined as:

$$\Delta P_{addv} = 3 \sum_{v=3}^{n} I_v^2 R_v, \qquad (1)$$

where I_v - the current of v-harmonic; R_v - active resistance of power lines at the frequency of v- harmonic.

In general, we recommend determining the active resistance of overhead power lines taking into account the function (Zhezhelenko 2012):

$$R_v = R_2 K_r K_{rv}, \qquad (2)$$

where $R_2 = r_0 l$ - the active negative (phase-) sequence resistance of the power lines; $K_r = \sqrt{v}$ - a coefficient which takes into account the dependence of active resistance on frequency; K_{rv} - the correction factor, which takes into account the distribution of the parameters in the equivalent circuit.

Additional energy loss is determined in accordance with the largest power loss (the simplified method recommended if there is reliable information about power consumption):

$$\Delta A = \Delta P_{addv} \cdot \tau_{max} \qquad (3)$$

where τ_{max} - the time of biggest loss, defined by the time of the heaviest loads T_{max}, which had been obtained in the analysis of real graphs of harmonic currents (daily $T_{max} = 15$ hours, annual $T_{max} = 6000$ hours).

The main regulatory document where the requirements to the quality of electrical energy in the power networks of general purpose are specified is GOST 13109-97. On 01.01.2013 Russia has adopted a new State Standard which takes into account the electromagnetic interference in the operation of electrical equipment of the new generation.

Under this Standard, a part of PQR describes EMF in steady conditions of equipment operation as far as electric power supplier and consumers are concerned due to the peculiarities of the process of production, transmission, distribution and consumption. These ones include voltage and frequency deviation, distortion of voltage waveform harmonicity, voltage unbalance and fluctuations. For their normalization the permissible PQR values have been determined.

Another part of PQR involves short-term EMF resulting from commutation processes, thunderstorm and other atmosphere phenomena as well as post-emergency modes: voltage drops and pulses, short interruptions in power supply. The Standard does not predetermine any acceptable numerical values for such cases.

PQ standards according to GOST 13109-97 are the levels of electromagnetic compatibility (EMC) for conducted electromagnetic interferences (EMI) in general-purpose power supply systems. Adherence to the standards provides EMC of general-purpose electric networks and power grids of consumers (receivers of electric energy).

Compliance with the power quality standards makes possible not only to save fuel and energy resources, but the other types of material resources as well, a part of which, at a reduced PQ, has to be spent on products rejected as defective or recyclable ones (Zhezhelenko 2004, 2012).

Reduction of upper harmonics levels (UH) in electrical networks is a part of the general task aimed at minimizing the influence of non-linear loads on the mains supply and at improving the quality of electricity supply in power systems of enterprises. Comprehensive solution to this problem based on the use of multi-function devices, is economically more appropriate than, for example, the use of measures to improve the network transformer current waveform. Resonant filters, known as compensating filter-devices (CFD), which, along with reduction in the levels of UH generate reactive power into supply network are the example of such multifunctional devices.

When connected in parallel LC-circuits are tuned to the frequencies of individual harmonics, resonant CFD works in practice. Deficit of reactive power in substation buses in this case can be completely compensated by means of CFD capacitor banks, moreover the installed power of capacitors is used by 80-90%. Thus, CFD are the simplest and most cost-effective filters, and this fact ensured their widespread use.

If the range of the power load in the shaft involves both power collectors and frequency converters, then unwanted (hazardous) resonance phenomena occur caused by the presence of an oscillatory circuit such as mains supply – static capacitor banks (SCB). The peculiarity of these modes is the correlation between the stationary electrical equipment units in the shaft (main fans, hoisting systems, transport) and technological cycles.

To analyze and evaluate resonance phenomena in power systems of mines, it is necessary to model frequency characteristics (frequency response FR) of electric circuit in view of mutual resistance (conductivity) of its individual components (off-take cables, transformers, motors). The resulting frequency responses will help identify the area where resonance frequencies appear when SCB of certain capacity is connected to the real-mode power supply system (the plant capacity, the number and parameters for off-take cables are taken into account).

Frequency response calculation algorithm has been worked out and tested by practical measurements. Initially, the simulation requires the most accurate information about the parameters of the power supply system of mines (length and cross section of overhead power lines, power transformers' capacity, types of current limiting reactors, SCB power (the resistance of the system verified by testing). Then electricity systems with typical (normal) mode of operation are chosen and regimes that are observed during repair-and-renovation operations or post-fault operative switching are identified. Next, a replacement scheme for each analyzed mode is made up and the resultant RLC network impedance and the load are determined by electric and technical calculations.

To construct a frequency response, one can use different mathematical approaches (experimental methods based on active or passive experiment and involving spectral analysis, the use of wavelet transformation for experimental identification, determination of network frequency responses using the correlation moments of currents and voltages). All these methods require active intervention in the existing power supply system of mines, so the principle of engineering calculation of frequency response is used in this research. If there are reliable baseline data, this approach will give acceptable results for the evaluation and development of technical recommendations to improve electrical systems.

The obtained frequency responses clearly show the areas of resonance occurrence and help take decisions whether to install SCB, active or hybrid filters, which methods for SCB protection to employ, etc.

3 FREQUENSY RESPONSE IN POWER SUPPLY SYSTEMS

For example, calculation of frequency response in power supply for mine "Blagodatnaya", at normal (each high voltage input works separately) and post-fault operation modes (only one input is enabled, the section switch is in the "on" position).

There are the basic elements involved in resonance as shown in the equivalent circuit (Fig. 1-b): the mains power supply, battery cosine capacitors and valve load (VLC with the converter).

The connection of capacitor banks results in a non-linear character of the frequency response (Fig. 2-3). Nonlinearity of the characteristics is determined by a Q element of the supply mains (the ratio x / r).

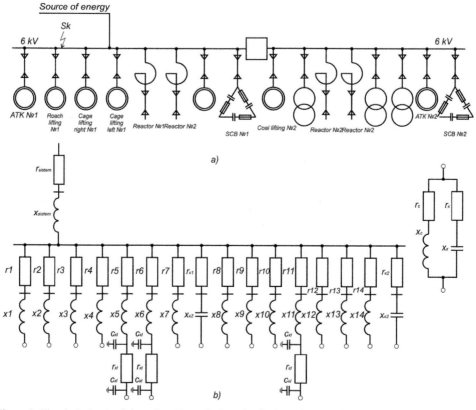

Figure 1. Electrical circuit of the nains (a), equivalent circuit (b) and the equivalent circuit for the post-fault operation of the mine

Nonlinearity of the frequency response of supply mains is explained by the fact that when the SCB is connected, parallel LC - circuit consisting of inductive reactance of supply mains and capacitive resistance of condenser is formed.

The equation for the frequency response of LC circuit of supply mains, forming the basis of the resonance phenomena model, is as follows:

$$Z_{1v} = \frac{\left(r_c + jx_{cv}\right)\left(r_k - jx_{kv}\right)}{r_c + r_k + j\left(x_{cv} - x_{kv}\right)} =$$

$$\frac{r_c r_k \left(r_c + r_k\right) + x_{cv}^2 r_k + x_{kv}^2 r_c}{\left(r_c + r_k\right)^2 + \left(x_{cv} - x_{kv}\right)^2} + \tag{4}$$

$$+ j \frac{x_{cv} x_{kv}\left(x_{kv} - x_{cv}\right) + r_k^2 x_c - r_c^2 x_k}{\left(r_c + r_k\right)^2 + \left(x_{cv} - x_{kv}\right)^2}$$

where x_{cv} - the equivalent inductive resistance of supply mains at v-harmonic frequency;

$x_{k\nu}$ - capacitive resistance of the capacitor bank for the ν- harmonic.

The equation has been processed in the mathematical package, and for the conditions of power supply circuit at mine " Blagodatnaya " the following values of the equivalent circuit elements have been obtained:

$r_k = 0,052\,\Omega$ - active resistance of circuit SCB1;

$r_c = 0,01\ \Omega$ - the total active resistance of the network elements;

$x_k = 22,9\ \Omega$ - nominal capacitive resistance SCB1 with capacitance C = 126.3 microfarad;

$x_c = 0,7\ \Omega$ - inductive resistance of supply mains at the basic frequency;

Capacitive resistance of capacitors is determined by the given reactive power $Q_{SCB1} = 1575$ volt-amperes reactive.

As a result of modeling, the frequency responses of the network $z = f(\nu, H_z)$ are obtained for different modes of mine operation (Fig. 2, 3).

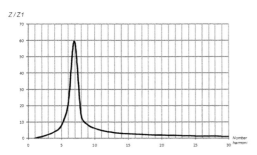

Figure 2. Type of frequency response for normal mine operation

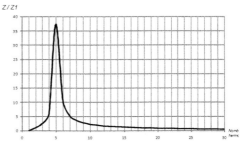

Figure 3. Type of frequency response for the post-fault mine operation

Thus, it is clear that in post-fault mine operation, the resonant frequency is reduced (the resonance frequency band is 200-350 Hz, the main resonance harmonic is 5). In emergency mode (parallel operation of two inputs), the resonance frequency is increased (the resonant frequency band of 300-400 Hz, the main resonance harmonic is 7). Such regimes lead to accidents and damage in the electrical system.

By studying the normal mine modes, we obtained frequency characteristics of supply mains at nonlinear loads. It was found out that when SCB is connected, frequency response is dramatically uneven. The band of frequencies where current resonance occurs in the range of 200-400 Hz. The main resonant harmonics are 6 and 7, respectively, for 1 and 2 bus sections.

Having analyzed frequency characteristics of the post-fault and emergency modes of power systems, which were obtained the model of resonance processes calculation, we could find out that the resonant frequencies of these modes are also in the range of 200-400 Hz, but the main resonant harmonics are 5 and 7. The calculation

results are completely consistent with experimental studies carried out at the mine. These findings prove the reliability of the results obtained.

Currently researchers have accumulated a lot of positive experience in dealing with a non-sinusoidal voltage in supply systems at coal mines with strong non-linear loads. In 2010, CFD 5 harmonics produced by "CKD ELECTROTECHNIKA" (Czech Republic) were installed at the "Geroiev Kosmosa" mine. A simplified diagram of CFD is shown in Figure 4. CFD is configured to filtering the 5th harmonic; power of capacitors is 2 × 2500 quarter. Prerequisites for their installation were: modernization of main fans, replacement of the drive motor, and installation of asynchronous thyristor cascade (ATC) to control the mode of ventilation. In addition, coal and rock hoisting is equipped with a DC motors with valvetype frequency converter and the control system EDPTS (electrical drive prepacked thyristor). All this has led to a sharp derating in terms of non-sinusoidal voltage.

15

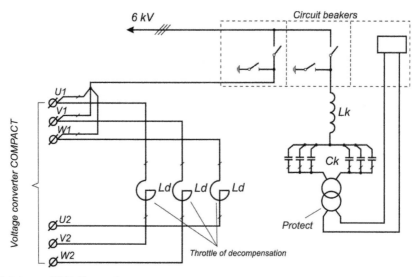

Figure 4. Scheme of CFD 5 harmonic

Taking into account that the mine "Geroiev Kosmosa" is the first mining enterprise in Ukraine which used adjustable CFD, an experimental analysis of voltage quality under different operating conditions of the mine was conducted. The purpose of the study was:

- to determine the levels of voltage waveform distortion when VLC work with ATC system;

- to determine the levels when the non-sinusoidal voltage coefficients and n- harmonic components decrease if CFD of 5th harmonic is used;

- to detect the influence of BSC capacity on K_U and $K_{U(n)}$ increase due to resonance effects in the network.

As a result of integrated practical studies conducted at the mine "Geroiev Kosmosa" with the help of voltage quality analyzer "FLUKE 435", the time dependences of non-sinusoidal quality (Fig. 5) and power factor (Fig. 6) were obtained. This relations show the effectiveness of PKU 5 harmonic.

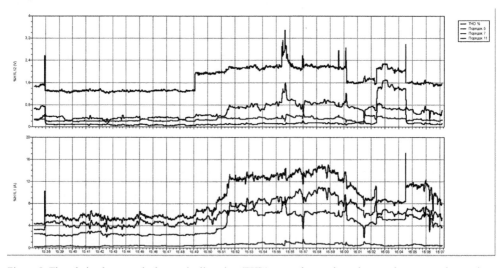

Figure 5. The relation between the harmonic distortion (THD), upper harmonics voltage and current and operation modes of the mine

16

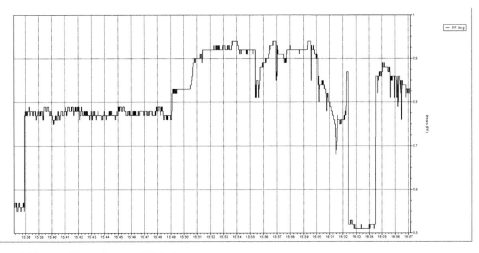

Figure 6. Change of the factor cosφ (PF) in time

Analyzing the dependence, we see that when CFD was switched at 15.38, the harmonic distortion coefficient decreased from 1.6 to 1.5%. Similarly, the level of the seventh harmonic decreased. The fifth harmonic is almost completely absorbed from the network. Small bursts of indices that were observed at 15.55 and 16.00 are due to manifestation of resonance phenomena in the oscillatory circuit "mains supply - static capacitors bank." During the period 16.02-16.04 CFD was off, and the increased level of performance has been immediately indicated this fact. Furthermore, the performance of CFD connected to compensate the reactive power in the electricity system of the mine should be noted. As the measurements testify, during the period from 15.38 to 16.02 the average power factor cosφ (PF) is in the range 0.85, which is an excellent result for a coal mine. However, when we disabled CFD from 16.02 to 16.04, the average PF decreased to 0.52 at a constant power load of the mine. Thus, the case studies of quality voltage have shown the effectiveness of adjustable CFD to improve power quality and power factor correction in networks of coal mines in Ukraine.

CONCLUSIONS:

- In the analysis of frequency characteristics for the power supply network in coal mining corporation "Pavlogradugol", the areas with resonance phenomena manifestation have been identified. The physical nature of these phenomena is explained by the nonlinear nature of frequency response in case the network has capacitor banks and frequency transformers/converters, which are becoming commonly used nowadays. The shortcomings of these processes are related to the possibility of SCB current overload and untimely wear of the capacitors;

- Experimental measurements taken at the mines showed that their electric circuits are loaded with reactive power (power factor in some modes is extremely small). Hence, it is necessary to take precautions in order to compensate the reactive power, proving the importance of installing BSC with error-free performance in supply systems. Therefore, the protective reactors must be installed to protect the battery against overload by currents of upper harmonics.

- Theoretical calculations of frequency response and experimental measurements of operation modes of mines have shown that resonance processes occur at all mines to a greater or lesser degree. It becomes obvious when BSC is connected and thus, the level of the higher harmonics of 250-350 Hz increases (resonates). The onset of resonances is inevitable in case BSC and frequency converters of various types are connected simultaneously. In order to avoid the occurrence of emergency operation of power supply systems at mines, to reduce additional energy loss, to extend the service life of the insulation of electrical machines and cables, it is necessary to control the levels of upper harmonics when converters are installed. The problem will acquire cutting-edge dimension after adoption of the new GOST (State Standard) in Ukraine applied to the quality of electricity following the development of the methods ensuring the division of responsibility for the shared ownership between utility companies and consumers in case of the voltage drop.

- The results of theoretical and practical studies of voltage quality indices demonstrate that the installation of powerful non-linear power collectors at coal mines leads to the systemic derating of these indices. At the same time, the effectiveness of the 5 harmonic CFD installation has been proved, and this is a universal solution to the problem of voltage quality and reactive power compensation. The technical feasibility of similar installations at other mining companies is conditioned by complex technical and economic assessment.

REFERENCES

GOST 13109-97. Electrical energy. Electromagnetic compatibility. Quality standards for power supply systems in general. - Introduced in Ukraine 01.01.2000. - (Interstate standard CIS):30.

Zhezhelenko, I.V. 2004. *Higher harmonics in the power systems of industrial enterprises* (in Russian). Moscow: Energoatomizdat: 358.

Zhezhelenko, I.V. at al. 2012. *EMC customers* (in Russian). Moscow: Mashinostroenie: 350.

European Committee for Electrotechnical Standartization. Electromagnetic compatibility (EMC). Generic standards. Emission standard for residential, commercial and light-industrial environments EN 61000-6-3: 2007.

Zharkin, A.F., Novskiy, V.A. & Palachev, S.A. 2010. *Laws and regulations affecting the quality of electric power. Analysis of Ukrainian and European legislation and normative documents* (in Russian). Kiev: Institute of Electrodynamics of NAS of Ukraine: 167.

Energy Efficiency Improvement of Geotechnical Systems – Pivnyak, Beshta & Alekseyev (eds)
© 2013 Taylor & Francis Group, London, ISBN 978-1-138-00126-8

Limitations of the indirect field oriented control utilization for electric drives of pipeline valves

O. Beshta, I. Yermolayev
State Higher Educational Institution "National Mining University", Dnipropetrovs'k, Ukraine

K.-H. Kayser, N. Neuberger
Esslingen University of Applied Sciences, Göppingen, Germany

ABSTRACT: The paper deals with issues of the vector control utilization for electric drives of the pipeline valves. Because of the design features of the pipeline valves, the usage of an encoder on the output shaft of the gearbox results in error in determining position of the rotor flux vector. This error leads to the limitations of the indirect field oriented control of the pipeline valve, which were determined using different encoder resolutions. The analysis of the research and the problems of vector control utilization of pipeline valve electric drives are discussed.

1 INTRODUCTION

An important step in reducing the costs of production with the rising prices of energy and raw materials for metal, machinery, chemicals, oil and gas industries is optimizing the management process of flows using the pipeline valves of primary importance here is the search for technological solutions to improve the quality of the production process by means of the drive, troubleshooting and prevention of failures of industrial plant valves.

Modern world trends of automated process control systems in the automation control valves determine the main directions for the development of "smart - valves", which have a wide range of electronic control and allow to evaluate their condition, which ensures their work in different modes of operation in the process.

One of the important criteria for the use of electric drive is the continuous monitoring of the torque that can detect changes in the pipeline valves (eg, inertia, wear, etc.) and to perform preventive maintenance (Kayser, Beshta & Yermolayev 2012).

Large part of the AC drives, used in the modern pipeline valves, is obsolete. The traditional solution for motor control valves is automatic relay, where the position of the shut-off element is controlled by limit switches; as well for the limitation of the torque claw or friction clutches are used. If the valve actuator operates with frequent starts, it uses Gate Turn-off (GTO) thyristor starters that do not have drawbacks inherent in electromagnetic starters, but their high power losses in the protection circuits and during switching also make them unreliable. Moreover, frequency converters with scalar control system have been widely used lately. This kind of drives has more efficient control and a significant reduction of the drive due to the refusal of the torque limiting clutches and switches. However, they do not provide the direct torque control; have low accuracy and poor dynamic performance (Antropov 2010, 2011; Garganeyev & Karakulov 2006).

None of these control systems fully satisfies requirements for pipeline valves set in (Kayser, Beshta & Yermolayev 2012). One solution for this problem is the use of the vector control system.

Let us consider the standard control system of the valve electric drive (Fig. 1). It consists of an induction motor, a worm gear (sometimes in combination with a planetary gear), optical or magnetic position sensor (absolute – multiturn encoder) and a valve itself. Compared to the standard control technique and the indirect field oriented control (Fig. 2), becomes obvious that the difference consists only in installation of an additional current sensor for measuring two phase currents of the motor. As an expensive part of the structure namely encoder is used in both schemes and software, which has to be developed only once, leads to a slight increase in the cost of the control system.

Vector control system operation is not possible without information of the rotor flux vector position at each point of time. Motors with built-in sensor of the magnetic flux are not used, so in these systems the magnetic flux vector must be determined by the position or the speed sensor and an observer.

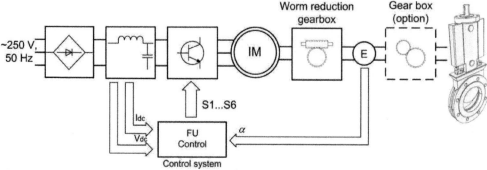

Figure 1. Block diagram of the FU control of the pipeline valve electric drive

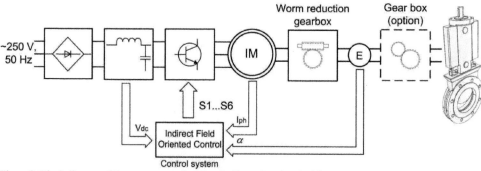

Figure 2. Block diagram of the vector control of the pipeline valve electric drive

Vector control actuators are sensitive to changes in the parameters of the drive. If the parameters of the controlled system, such as resistance and inductance of the rotor, are set incorrectly and the quality of the current and encoder measurements are low, flux observer determines its magnitude and position with an error. It is noted that the error in determining parameters of the induction motor leads to a significant degradation in the quality of regulation, and may lead to loss of stability.

Thus, in the system "induction motor - gearbox - encoder - pipeline valve" the problem of determination of the rotor flux vector position of induction motor is vital. Since the accuracy of the speed and position of the input shaft of the gearbox is inversely proportional to the gear ratio, which leads to the determination of the rotor flux vector only at certain times, such a method of measuring the position leads to a large control error and makes the vector control applicable only with restrictions (Kayser, Beshta & Yermolayev 2012).

Based on the above, the main idea of this paper is to determine the limitations of the indirect field oriented control of induction motor using different encoder resolutions and to make the analysis of its utilization.

The research can be approached in different ways:

- theoretical method, in which a test system is described as a mathematical model. In this case it is necessary to make a number of assumptions in the description of differential equations, the extent of which does not allow a qualitative assessment of the process;

- simulation of the system by using simulating software which can allow to take into account certain processes, neglected by theoretical methods. For example, simulation of current sensors and encoder signals in the presence of noise, as well as the sampling and quantization effects. However, if it is a significant complication of the model which leads to more

20

reliable results, the final stage of all the designs of complex nonlinear control systems, such as indirect field oriented control, should be experimental research;

- experimental approach, which is using concrete implementation of the control system, gives the most realistic results, because at the level of the mathematical models is very difficult to consider all the disturbing factors that are neglected in the analytical method;

In the research, theoretical consideration were made, which were then further tested practically on the experimental sample

2 MODEL DESCRIPTION AND CONTROL DESIGN

2.1 *Dynamic model of induction motor in the field coordinates*

Three phase squirrel cage induction motor in synchronously rotating reference frame can be represented as Fig. 3 (Bose 2001)

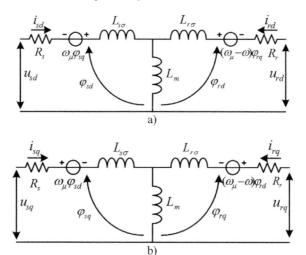

Figure 3. Equivalent circuit of induction motor in synchronous rotating reference frame; a) q-axis circuit b) d-axis circuit

The induction motor model, described by using a space vector notation and written in (d, q) reference frame, rotating with the synchronous angular velocity ω_μ, is presented in the following equations (Leonhard 2001):

$$u_{sd} = i_{sd}R_s + \frac{d\varphi_{sd}}{dt} - \omega_\mu \varphi_{sq} \qquad (1)$$

$$u_{sq} = i_{sq}R_s + \frac{d\varphi_{sq}}{dt} + \omega_\mu \varphi_{sd} \qquad (2)$$

$$u_{rd} = i_{rd}R_s + \frac{d\varphi_{rd}}{dt} - \omega_r \varphi_{rq} \qquad (3)$$

$$u_{rq} = i_{rq}R_s + \frac{d\varphi_{rq}}{dt} + \omega_r \varphi_{rd} \qquad (4)$$

where u_{sd} and u_{sq} are the (d, q) axis stator voltages, (i_{sd}, i_{sq}) and (i_{rd}, i_{rq}) are respectively the (d, q) axis stator and rotor currents, ω_μ and ω_r are respectively the reference frame and the slip angular velocity; $(\varphi_{sd}, \varphi_{sq})$ and $(\varphi_{rd}, \varphi_{rq})$ represent the (d, q) axis stator and rotor fluxes. They can be described by the following equations:

$$\varphi_{sd} = L_s i_{sd} + L_m i_{rd} \qquad (5)$$

$$\varphi_{sq} = L_s i_{sq} + L_m i_{rq} \qquad (6)$$

$$\varphi_{rd} = L_r i_{rd} + L_m i_{sd} \qquad (7)$$

$$\varphi_{rq} = L_r i_{rq} + L_m i_{sq} \qquad (8)$$

21

where L_s and L_r are stator and rotor inductance, L_m is the mutual inductance. The mechanical and the electromagnetic torque equations are as follows:

$$T_e - T_l = J\frac{d\omega}{dt} \tag{9}$$

$$T_e = \frac{Z_p L_m}{L_r}(\varphi_{rd}i_{sq} - \varphi_{rq}i_{sd}) \tag{10}$$

where T_e is the electromagnetic torque, T_l is the load torque, ω is the rotor angular velocity, J represents the total inertia and Z_p is the number of pole pairs.

2.2 The indirect field oriented control (IFOC)

According to the rotor field oriented control theory (Blaschke 1972; Trzynadlowski 2001), the stator current of the induction motor can be decomposed into two orthogonal components in the synchronous rotating reference frame, which are responsible for the torque (q axis) and the rotor flux (d axis) generation. The purpose is to independently control the torque and the flux such as in a separately excited direct current (DC) machine. Fig. 4 shows this orientation:

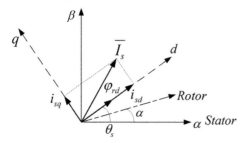

Figure 4. Reference frame and space vector representation

where θ_s and α are respectively (d, q) the reference frame and the rotor position. Considering the orientation in Fig. 4 we have:

$$\varphi_{rd} = L_m i_{\mu d} \text{ and } \varphi_{rq} = 0 \tag{11}$$

where $i_{\mu d}$ represents magnetizing current according to the axis d.

From the induction motor equations (1) – (10) and considering the result of the orientation given by (11), we have:

$$i_{sd} = T_r\frac{di_{\mu d}}{dt} + i_{\mu d} \tag{12}$$

$$\frac{1}{T_r}i_{sq} = \omega_r i_{\mu d} \tag{13}$$

where T_r represents the rotor time constant.

From equation (12), the amplitude of the rotor flux linkage can be maintained at a fixed level by controlling the direct-axis stator current, (12) in steady state can be written as:

$$i_{sd} = \frac{i_{\mu d}}{L_m} \rightarrow i^*_{sd} = \frac{i_{\mu d}}{L_m} \tag{14}$$

where * represents the reference value.

In (13) the term $\dfrac{i_{sq}}{T_r i_{\mu d}} = \omega_r$ represents the slip angular velocity of the rotor flux ω_r, considering that the synchronous angular velocity is equal to the sum of the electrical rotor angular velocity and the slip angular velocity of the rotor flux. Thus, we obtain:

$$\theta_s = \int \omega_\mu dt = \int (\omega + \omega_r)dt = \alpha + \frac{i_{sq}}{T_r i_{\mu d}} \tag{15}$$

According to the equation (14), the division of the direct and quadrature - axis stator currents (i_{sd}, i_{sq}) depends on the slip angular velocity ω_r, and the two feedback current signals are used to determine the required slip angular velocity. When the rotor position and the value of the slip angle are added, the rotor flux position is obtained.

Equation (10) shows that the electromagnetic torque expression in the dynamic mode presents an interaction between the rotor flux and the stator current. In case of the indirect field oriented control, the electromagnetic torque can be expressed as:

$$T_e = \frac{Z_p L_m^2}{L_r}i_{\mu d}i_{sq} \tag{16}$$

Hence

$$i^*_{sq} = \frac{Z_p L_m}{L_r i^*_{\mu d}}T^*_e \tag{17}$$

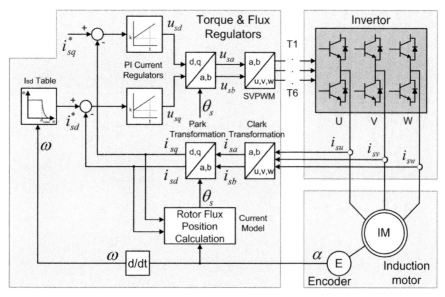

Figure 5. Block diagram presenting the model of the indirect field oriented control of induction motor

Considering a complete decoupling of the torque and flux, the electromagnetic torque generated by the motor can be controlled by controlling the q-axis current, if the rotor flux is kept constant.

Also from the state model, we have:

$$i_{sd} = (\frac{u_{sd}}{R_s} - (1-\sigma)T_s \frac{di_{\mu d}}{dt} +$$

$$+ \sigma T_s \omega_\mu i_{sq})_r \frac{1}{1+\sigma T_s s} \qquad (18)$$

$$i_{sq} = (\frac{u_{sd}}{R_s} - (1-\sigma)T_s \omega_\mu i_{\mu d} -$$

$$- \sigma T_s \omega_\mu i_{sd}) \frac{1}{1+\sigma T_s s} \qquad (19)$$

Control loops of the d and q axis of the stator current are internal closed loops and are crucial to the dynamics of the electric drive based on induction motor. The dynamics of the stator currents is represented by simple linear first order differential equations. That means that a simple PI controller can be used to ensure currents components control. $W_{sd}(s)$ and $W_{sq}(s)$ represent the d and q axis transfer functions of the induction motor. They are given by:

$$W_{sd}(s) = W_{sq}(s) = \frac{1/R_s}{1+\sigma T_s s} \qquad (20)$$

where s denotes Laplace variable.

The delay caused by a large part of the stator time constant T_s, is compensated by the boost component of the regulator, and the gain is selected by the principle of "optimum modular", which corresponds to the damping coefficient of the oscillating transfer function $D = 1/\sqrt{2}$.

Fig. 5 shows the block diagram of the proposed indirect field oriented control model of induction motor. The actual stator currents i_{su}, i_{sv}, i_{sw} are detected using the current sensors. The current control loops receive the torque and excitation current commands i_{sq}^*, i_{sd}^* and the actual current i_{sq}, i_{sd}. Simple PI controllers are used for both i_{sq}, i_{sd} current loops. A Space Vector PWM is used to apply the voltage commands.

2.3 Theoretical considerations on the effect of the encoder resolution impact on the vector control

Consider the induction motor in a form of the generalized machine with one pole pair. In this case the failure in the orientation must be less than $\gamma = 90°$, otherwise the control of the motor

23

is no more possible. Thus all the subsequent reflections will be considered in one quadrant (Fig. 6a).

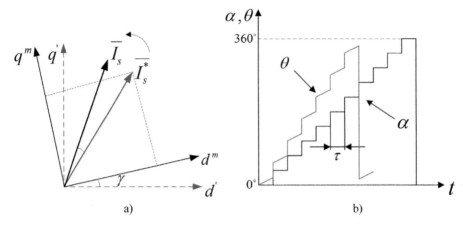

a)   b)

Figure 6. a) Reference frame and space vector representation in case of the field orientation failure ($d^m q^m$ - motor reference frame, $d' q'$ - reference frame of the control system) b) Relation of the rotor flux angle θ to rotor angle α

In the absence of identification error of the rotor flux vector position, the condition of ideal orientation of the rotor flux vector ($\gamma = 0°$) takes place and the feedback signals are exactly the true values of the induction motor currents. However, the lack of the information about the motor shaft position results in the error of calculation of the rotor flux vector phase. In this case all feedback signals measured by the current sensors are no longer correct, as can be seen from Fig. 6a.

Two systems of d, q coordinates can be determined in such situation in the electric drive with vector control. One system is aligned along the true rotor flux vector ($\varphi_{rq} = 0$), and the other - along the observed vector ($\varphi_{rq}^* = 0$). The transition between these coordinate systems can be described as a shift in the difference of the rotation angle of the true and the observed vectors. Obviously, when $\varphi_{rq} \neq 0$, feedback signals do not correspond to the true values of the induction motor, which leads to a static control error as well as to the occurrence of a non-deterministic dynamics of the system, including the loss of system stability.

The information about the motor shaft position depends on the speed and quantization effect (encoder resolution) of the rotor angle (Fig 6b). As stator and rotor time constants also produce an impact on the error in the field orientation when these constants have smaller value than the time till arrival of the next angle information (T_s and / or $T_r < \tau$), the feedback of the current signals can be already changed according to the control response. So more clearly the error in the field orientation becomes more apparent at low speeds, because information about the motor shaft position is unknown for the period of time that is greater than electrical time constant. At high speeds, the time to update the position of the rotor is much smaller than the electric time constant, which accordingly does not lead to a torque ripple. However, at zero speed the measured torque corresponds to the set one, as there is no incoming information about the rotor angle and the rotor flux orientation depends entirely on the slip angular velocity.

Orientation failure, when $\gamma \neq 0°$, leads to the loss of stability of the vector control. The process of the stability loss is accompanied by the motor demagnetization or saturation of the magnetic circuit, leading to decrease of the motor torque or to higher losses related to increase of the current. At the same time the scenario of the motor dynamics is defined according to the sign of the torque component of the stator current and the error of the angle γ.

Let us assume that rated flux component $i_{sd\_rated} = 1$ A and rated torque component $i_{sq\_rated} = 2,5$ A. In case of inconsistency of true and observed coordinate systems and absence of saturation effect and voltage limitations, the d-

24

axis current increases, and the q-axis current decreases according to Fig. 7. When the drive operates at a low speed, and the slip angular velocity is larger compared to the rotor one, the field orientation mainly depends on the value of the slip angle. But when the new information of the encoder comes, the observed rotor flux position corrects itself which results in the change of the currents feedback signals and controllers' outputs respectively. That in turn could lead to saturation effects.

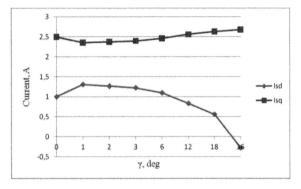

Figure 7. Relation between currents i_{sd}, i_{sq} variation and the failure in the field orientation

3 IMPLEMENTATION AND RESULTS OF INDIRECT FIELD ORIENTED CONTROL OF INDUCTION MOTOR

3.1 *Experimental Setup*

A three phase induction motor with a rotor flux oriented vector control was used for this research. Rating and the parameters of the motor are given in Table 1. The experimental setup (Fig. 8) consists of a tested induction motor fed by a voltage source PWM inverter, regulated by the vector control and a DC generator, which is mechanically coupled to the motor for the load regulation, having the following characteristics: 32 W, 24 V, 2.5 A, 1300 rpm.

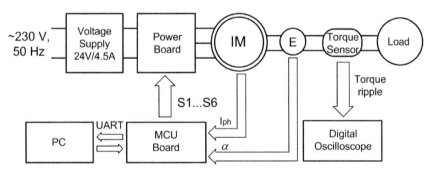

Figure 8. Block diagram of the experimental setup

The flux position estimator (current model), the current controllers, and the coordinate transformations were build using Toshiba Microcontroller (TMPM374FWUG), and the pulse width modulations of the stator voltages were made by Toshiba Power Board M374PWR, where the DC link voltage and the carrier frequency of the PWM inverter were set to 24V and 10kHz respectively. The setup also comprises two current sensors (ACS712) that are required for the field oriented control structure and Siemens incremental encoder 6FX – 2001 with a resolution of 1000 increments per revolution used for estimating rotor shaft speed and position.

Table 1. Parameters of the tested Induction Motor

Components	Parameters	Values
P	Rated Power	0.02 kW
n	Rated Speed	900 rpm
I	Rated Current	4.2 A
U	Rated Voltage	24 V
R_s	Stator Resistance	1.8 Ohm
L_s	Stator Inductance	0.016 H
T_r	Rotor Time Constant	18 ms
Z_p	Number of pole pairs	2

3.2 Experimental Results

In order to analyze the drive system performance, the research is carried out into the induction motor drive taking into account the variation of the encoder resolution. The drive was operated in the torque stabilization mode at different speeds. The results were shown at lowest speeds with an encoder resolution of 16 – 40 incr/rev (which corresponds to 8-20 incr / electrical rev). The

torque measurements were made in AC measurement mode, so only torque ripple is presented in the Figures.

Fig. 9 and Fig. 10 show the effect of the encoder resolution <u>impact</u> on the torque. It can be seen that when the speed is reduced, the amplitude of the pulsation, relative to the set torque, is increasing (approximately by factor 2). Taking into consideration results of Fig. 11, increase in the encoder resolution to 40 incr/rev at the same speed leads to less amplitude of the torque ripple.

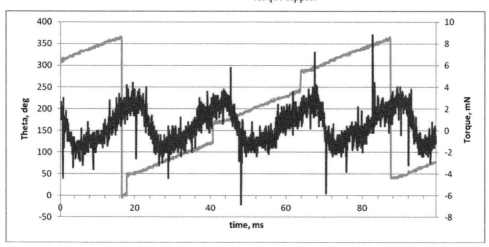

Figure 9. IFOC performance with 16 incr/rev at 160 rpm and 50% of the rated torque: grey - rotor flux angle, black – torque ripple

26

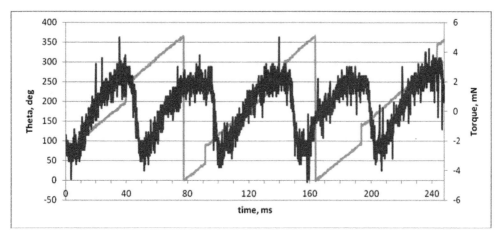

Figure 10. IFOC performance with 16 incr/rev at 80 rpm and 50% of the rated torque: r grey - rotor flux angle, black – torque ripple

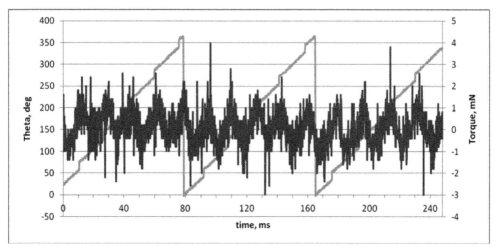

Figure 11. IFOC performance with 40 incr/rev at 80 rpm and 50% of the rated torque: grey - rotor flux angle, black – torque ripple

4 CONCLUSIONS

Analysis of the results shows that deterioration of the encoder resolution at low speeds leads to the periodic field orientation failure which in case of time constants, smaller than the time of the encoder information updating, can affect the stability of the electric drive system.

Thus, to maximize the use of the induction motor torque at zero speed up to rated in indirect field oriented control of the pipeline valve, the future research will be aimed to determine the rotor flux vector position using:

1) extrapolation of the rotor flux vector angle or extrapolation with follow-up control;
2) standard sensorless vector control of induction motor with absolute multi-turn encoder subsequent correction;
3) information about the position of the rotor flux vector from sensorless vector control and position of the motor shaft from the encoder goes to smart filters (e.g. Extended Kalman Filter, Luenberger observer, etc.) which acts as an observer. The obtained approximated value of the filter is the rotor flux vector position.

REFERENCES

Kayser, K.-H., Beshta, A. & Yermolayev, I. 2012. *Current State and Development Prospects of Electric Drive of Pipeline Valves* (in Ukrainian). Thematic issue "Problems of automatic electric drive. Theory and application", Issue 3: 125-127.

Antropov, A. 2011. *Intelligent Electric Drives of Industrial Valves. Features of the application* (in Russian). «Armaturostroeniya», Issue 73: 69-72.

Garganeyev, A. & Karakulov, A. 2006. *Smart Electric Drive as Part of the Distribution of ACS* (in Russian). «Itech», Issue 4: 25-32.

Antropov, A. 2010. *Pipeline Valves. Improving Reliability of the Operation* (in Russian). «Itech», Issue 16: 26-30.

Bose, B. 2001. *Modern Power Electronics and AC Drives.* USA: Prentice Hall PTR: 711.

Leonhard, W. 2001. *Control of Electrical Drives.* Germany: Springer: 470.

Blaschke, F. 1972. *The Principles of Field Orientation as Applied to the New Transvector Closed Loop control System for Rotating Field Machines.* Germany: Siemens Review, Issue 34: 217-220.

Trzynadlowski, A. 2001. *Control of Induction Motor.* USA: Academic Press: 228.

Energy Efficiency Improvement of Geotechnical Systems – Pivnyak, Beshta & Alekseyev (eds)
© 2013 Taylor & Francis Group, London, ISBN 978-1-138-00126-8

Design of electromechanical system for parallel hybrid electric vehicle

O. Beshta, A. Balakhontsev & A. Albu
State Higher Educational Institution "National Mining University", Dnipropetrovs'k, Ukraine

ABSTRACT: The article deals with design of configuration and control system for hybrid electric vehicle of parallel topology. The demands for power train are formulated. It is shown that speed-torque characteristics of the internal combustion engine and the electric drive must be adjusted to provide best efficiency under parallel operation. Mathematical description and simulation of internal combustion engine are given. Measures for fuel economy are proposed.

1 INTRODUCTION

Hybrid electric vehicles (HEVs) are being developed and manufactured by almost every automobile company. Their benefits are well known: lower fuel consumption and thus less emission, higher dynamic performances. Nevertheless high initial cost of such vehicles and immature infrastructure restrict application of these vehicles among average customers.

There is an ongoing project at the National Mining University (Dnipropetrovs'k, Ukraine) dedicated to development of low-cost solution to retrofit conventional vehicle into hybrid one. Similar ideas were investigated by several researchers and engineers, like Bharat Forge and KPIT Cummins Infosystems LTD alliance which even announced that a special retrofit kit would be available on the market by 2011. The solution we pursue is to equip a front-wheel drive vehicle with additional electric drive installed onto the rear axle. The rear wheels can be equipped with hub motors, which requires less mechanical transformations and saves space but is more expensive, or the axle can be transformed to be driven from single electric motor. The latter solution despite its complexity is obviously cheaper and thus our target.

Since both motors are mechanically coupled via the road, we are dealing with parallel configuration of hybrid electric vehicle. Such configuration is known to provide better power/weight ratio for the vehicle. The electric motor can be of less power than the internal combustion engine, it can be used only as an auxiliary drive to fill dips in engine's speed/torque characteristic and to transform mechanical energy into electrical during braking. Despite all advantages of parallel topology the main obstacle is simultaneous operation of internal combustion engine and electric motor. In this article we shall discuss issues of parallel operation and consider the structure of control system for electric drive of parallel hybrid vehicle.

2 TOPOLOGY OF VEHICLE DRIVETRAIN AND ITS CONTROL SYSTEM

Let us consider the scheme of hybrid electric vehicle's powertrain shown in Fig. 1. The system contains componentsof power transmission: electric motor, internal combustion engine (ICE), torque coupler; and control modules for ICE and electric motor and the whole vehicle(Bogdanov 2009).

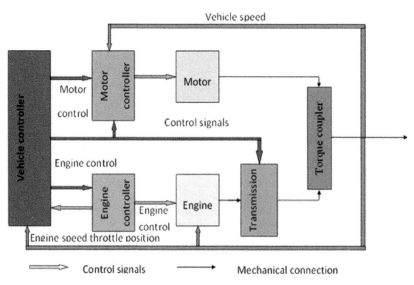

Figure 1.Conceptual architecture of a hybrid electric drive train.

The control system has two levels. The higher-level module is designed to manage overall vehicle by means of producing and distributing commands between lower-level control modules – ICE control module and electric motor control module. The control signals are formed according to certain control strategy which aims to fuel efficiency or dynamic performances. The reference signalis power demand from the driver, feedbacks include signals from the engine, motor and transmission.

In parallel configuration the total driving effort is produced by both primary source, which is internal combustion engine, and electric motor.

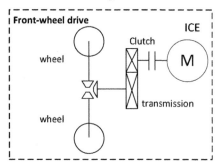

Figure 2.Configuration of parallel HEV drivetrain.

The torque coupling device that unites power flows from the engine and the motor is the factor that differs parallel configuration from the series one (Ehsani & Gao 2010). This element may or may not be present as a separate device. In our case the road itself is a torque coupler.

The interconnection of the engine and the motor via the road surface can be described mathematically as integration of the sum of torques, as it is shown in Fig.3.

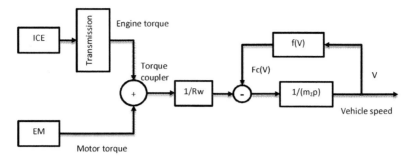

Figure 3. Mathematical representation of a torque coupling.

The rigid mechanical connection of the engine and the motor means that their speeds will always be maintained at certain ratio. Or, if reduced to wheel shafts, speeds of the engine and the motor will always be equal. The resulting speed-torque characteristic is obtained by summation of corresponding torques at certain speed levels.

Meanwhile there is always a danger of combining two motors with different speed-torque characteristics for parallel operation. The speed-torques characteristics always have different stiffness and/or no-load speeds. The motor with stiffer curve takes bigger part of the torque and can be overloaded. In worst cases when one of the motors has less value of no-load speed, it can be driven by its "partner" into the braking mode(Kolb& Kolb 2006).

An internal combustion engine is basically a torque source, it has soft speed-torque characteristic and little overload capacity. This being so, special measures must be taken for proper distribution of driving efforts between the motors and particularly to prevent operation of one of them in braking mode. It can be implemented by limitation of maximal speed reference for electric drive as it is shown in Fig. 4.

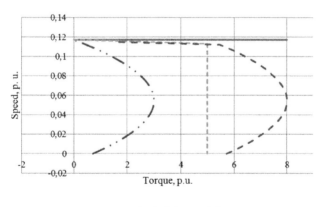

——· ICE mechanical characteristic

—— Speed limit

‐ ‐ ‐ ‐ speed-torque characteristic of electric drive

‐ ‐ Mechanical characteristic of HEV drive

Figure 4. Necessary shape of electric drive speed-torque characteristic.

Thus the speed-torque characteristic of electric part of the drivetrain must be absolute by soft in the range of speeds up to the value of no-load speed of the engine. Then it transforms into typical stiff characteristic of electric drive.

The maximum speed threshold must be a function of reduced to wheels value of ICE no-load speed. The latter value, in turn, depends on throttle position and current gear ratio. The higher the gear the greater the speed limit and the less is the current limit. At low gears the current is limited by natural overload capacity of the motor, at higher gears the motor produces less torque due to field weakening.

The key point in successful operation of parallel HEV is estimation of internal combustion engine's speed-torque characteristic. The position of operating point on this characteristic must be continuously tracked. The electric drive being more flexible and faster element can adapt to current conditions and drive the powertrain into point with higher performances.

Let us consider the performances of internal combustion engines.

3 DESCRIPTION OF INTERNAL COMBUSTION ENGINE

Interesting enough that first mechanically driven vehicles were electric ones. Performances of both types of engines – internal combustion and electric ones were improved with time and now they have roughly equal specific power. ICEs are known to have many disadvantages: high noise, bad speed-torque characteristic with dips, low overload capacity, and, of course, low efficiency and harmful emissions. Internal combustion engines got their chance only because of availability of hydrocarbon fuels and sufficient drive range they could provide at the dawn of automobile industry.

Nevertheless combustion engines in automotive applications will prevail in the near future. Let us consider mathematical description of the internal combustion engine. This will help when developing control algorithms for electric part of the HEV.

The model of ICE can be derived from mathematical description of processes within the engine. The model contains two blocks: one describes gas dynamics and the other describes mechanical part(Lamberson2008).

The air flow rate in the manifold system m_a is a function of pressure in p_m and throttle position θ:

$$m_a^{'} = f(\theta) \cdot g(p_m) \tag{1}$$

Each of the given elements can be presented as

$$f(\theta) = k_{th0} + k_{th1} \cdot \theta + k_{th2} \cdot \theta^2 + k_{th3} \cdot \theta^3 \tag{2}$$

$$g(p_m) = \begin{cases} 1, & p_m \leq 0.5 \cdot p_{atm} \\ \dfrac{2}{p_{atm}} \cdot \sqrt{p_{atm} \cdot p_m - p_m^2}, & p_m > 0.5 \cdot p_{atm} \end{cases} \tag{3}$$

where $k_{th0...3}$ – equation constants; θ – throttle position; p_{atm} – atmospheric pressure; p_m – manifold pressure.

The gas dynamics inside the intake system is described by first order differential equation:

$$p_m^{'} = \frac{R \cdot T_m}{V_m} \cdot (m_{ai}^{'} - m_{ao}^{'}) \tag{4}$$

Where R – gas constant; V_m – manifold volume; T_m – gas temperature in the manifold system.

The mass flow rate of the air going to combustion chambers from the manifold m_{ao} is a function of manifold pressure p_m and engine speed:

$$m_{ao}^{'} = k_{mo0} + k_{mo1} \cdot n \cdot p_m + + k_{mo2} \cdot n \cdot p_m^2 + k_{mo3} \cdot n^2 \cdot p_m \tag{5}$$

where $k_{mo0...3}$ – equation constants; n – engine speed.

The crankshaft dynamics is described by the equation

$$J \cdot n' = T_{eng} - T_l \tag{6}$$

Where T_{eng} – engine torque; T_l – load torque; J – engine inertia.

There is an empirical function for the engine torque

$$T_{eng} = k_{eo} + k_{e1} \cdot m_a + k_{e2} \cdot (AFR) + + k_{e3} \cdot (AFR)^2 + k_{e4} \cdot \sigma + + k_{e5} \cdot \sigma^2 + k_{e6} \cdot n + k_{e7} \cdot n^2 + + k_{e8} \cdot n \cdot \sigma + k_{e9} \cdot \sigma \cdot m_a + k_{e10} \cdot \sigma^2 \cdot m_a \tag{7}$$

where $k_{e0...10}$ – equation constants; m_a – mass of the air in the chamber; AFR – air-to-fuel ratio; σ – ignition advance.

The variable m_a represents air flow to the chambers during the intake. The intake takes place within first π radians of four crankshaft cycles. Thus the value of m_a may be obtained by integration of airflow from the manifold with zeroing it by the end of each cycle. To simplify representation of the engine, this process can be described as delay element (Khan, Spurgeony & Pulestonz 2008).

An integrator with variable zeroing time can be approximately described as

$$m_a = \frac{m_{ao} \cdot \pi}{n} \tag{8}$$

32

Where m_a – intake airflow; m_{a0}–output airflow; n – crankshaft speed.

The structure of internal combustion enginemathematical model is given in Fig. 5.

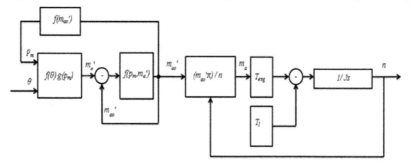

Figure 5. The structure of mathematical model of internal combustion engine.

The structure obtained can be used for design of control algorithms and simulation of vehicle dynamics. For preliminary configuration of HEV's components it can be useful to consider the resulting speed-torque characteristics of the internal combustion engine. The simulation of typical ICE installed in mid-class sedan gives the family of characteristics shown in Fig. 6.

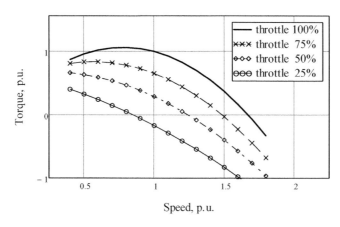

Figure 6. Speed-torque characteristics of internal combustion engine (in per units).

The family of characteristics is built for several positions of throttle, expressed in percents.

The obtained curves may be described by empirical equations such as

$$T(\omega,\theta) = k(\omega) \cdot T[\omega + \Delta\omega(\theta)] - \Delta T(\theta) \qquad (9)$$

Where $k(\omega)$, $\Delta\omega(\theta)$ and $\Delta T(\theta)$ – equation constants that are function of crankshaft speed and throttle position.

The key dependence is described by polynomial function such as

$$T(\omega) = -1.33 \cdot \omega^2 + 2.07 \cdot \omega + 0.251. \qquad (10)$$

Coefficients in the equation given were obtained for the engine of ZAZ Sens vehicle – low cost Ukrainian sedan.

4 SIMULATIONS AND FURTHER STUDIES

The above given mathematical descriptions were used in the complex model of parallel HEV. For electric drive, a simple cascade multiloop structure was implemented. The only distinctive feature in electrical part of the model was variable saturation of speed controller. The saturation function was set to coincide with no-load speed of ICE.

Figure 7 shows standalone operation of the ICE and Figure 8 shows transients in the vehicle during its acceleration on the first gear.

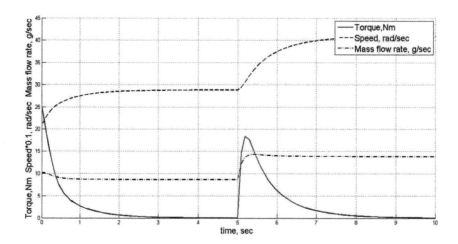

Figure 7.Simulation of internal combustion engine.

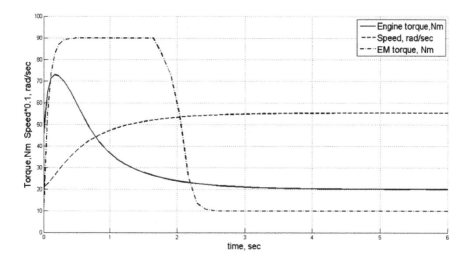

Figure 8.Transients in parallel HEV.

Opening the throttle causes the increase of the airflow, rise of engine torque and speed. Figure 8 shows efficient simultaneous operation of the ICE and the motor – neither of them produces braking torque.

So, the models given above can serve the basis for development of control strategies. The efforts should be directed onto optimization of operating modes of internal combustion engine while meeting demands for vehicle dynamics. Operation efficiency implies providing necessary torque by electric motor when ICE has unstable operation and driving the engine operating point

into areas with highest possible efficiency (Ehsani & Gao 2010).

Simultaneous operation is especially important during the initial phase of acceleration when ICE works unstable and with poor efficiency. At this stage the electric motor can even be overloaded to the limit of its capacity.

During continuous drive at cruise speed, as a rule, the power demand for the powertrain lies far below optimal operating point. In this case the so called "thermostat" or "on/off control strategy" may be applied for fuel economy. It implies periodic switching between two states: 1) operation in pure electric mode, when energy

34

is supplied from the battery; 2) ICE delivers energy for traction and for battery charging.

Capturing braking energy is a main source of fuel economy. In the braking mode, the electric motor operates as generator, delivering at least part of the energy to the battery.

5 CONCLUSIONS

Parallel configuration of hybrid electric vehicles will keep providing the best performance/cost ratio in the near-term future. The electric part of the drivetrain is flexible and fast enough to drive the internal combustion engine to the necessary operating point while meeting demands for tractive effort. The key issue in control of electric drive is correct distribution of torques between the engine and the motor. To do so, it is necessary to limit maximum speed of electric motor. The threshold should depend on instantaneous value of no-load speed of the engine reduced to the wheels. It, in turn, depends on current gear and throttleposition.

The model of the internal combustion engine can be used for research of vehicle dynamics and developments of electric drive control system.

REFERENCES

Ehsani, M. & Gao, Y. 2010.*Modern electric hybrid electric and fuel cell vehicles: fundamentals, theory and design.* Second edition. New York: CRC PressTaylor & Francis Group: 558.

Lamberson D. 2003. *Torque management of gasoline engines.*Available at s3.amazonaws.com/zanran_storage/vehicle.me. berkeley.edu/ContentPages/16150880.pdf.

Kolb, A. & Kolb, A. 2006. *Theory of electric drive.* Dnipropetrovs'k: National Mining University: 511.

Husain, I. 2010. *Electric and hybrid vehicles. Design fundamentals.* New York: CRC PressTaylor & Francis Group: 270.

Khan, M., Spurgeony, S. & Pulestonz, P. 2001. *Robust speed control of an automotive engine using second order sliding modes.* Proceedings of the European Control Conference, Vol. 5. Available at www.geocities.ws/khalid2k1/ecc2014.pdf.

Bogdanov, K. 2009. *Traction electric drive for automobiles.* Moscow: Higher School press: 57.

Energy Efficiency Improvement of Geotechnical Systems – Pivnyak, Beshta & Alekseyev (eds)
© 2013 Taylor & Francis Group, London, ISBN 978-1-138-00126-8

Cross- border cooperation of energy service companies as a factor enhancing energy and economic safety

O. Novoseltsev
NAS of Ukraine, Institute of Engineering Thermophysics, Kyiv, Ukraine

O. Kovalko
Gas Ukraine State Company, Kyiv, Ukraine

T. Evtukhova
Institute of General Energy, Kyiv, Ukraine

ABSTRACT: A comparative analysis of energy and economic efficiency and productivity of gross domestic product (GDP) in different countries is carried out. The conditions of comparative advantages and a procedure for cross-border cooperation of Energy Service Companies (ESCOs) to ensure profitability are formalized on the principles of performance-contracting. The factors of energy-economic interaction are defined in terms of client-alternative remuneration of ESCO services' customer, considering payments to ESCO under performance contracts, and the client-alternative costs. The results of numerical calculations that confirm the efficiency and effectiveness of the proposed approach for improving the energy-economic security are presented.

1 INTRODUCTION

Persistent growth of energy consumption in the world may be explained by the objective needs of economic development on the one hand, and limited or irregular positioning of natural ore and energy resources on the planet on the other hand. Hence extraction of natural resources, efficiency of their transformation and utilization rank first among the problems of economic, energy and ecological safety. Today these issues have to be tackled in conditions of active and even aggressive energy markets redistribution, their globalization and diversification of supply sources, use of nuclear energy and alternative energy sources, mitigating the impact of energy on the environment.

The level of Ukraine's energy security has abruptly deteriorated since 1991 as a result of the USSR disintegration, as Ukraine - unlike the former Soviet republics such as Azerbaijan, Kazakhstan, Russia, Uzbekistan and Turkmenistan - is not self-sufficient in energy resources. The situation is complicated by the uneven location of generation capacities, factories producing energy equipment and repair facilities. Thus, Ukraine has excessive potential of basic electricity capacities, while the maneuvering capacities, mainly hydropower, are situated mostly in Russia and Kyrgyzstan.

The following parameters are of primary importance among the main factors enhancing energy-economic safety: reliability of energy supply to industry and population, timely substitution of exhaustible energy resources by alternative ones, diverse kinds of fuel and energy, prevention of their inefficient use, incorporating requirements of ecological safety and environmental protection, utilization of industrial wastes as alternative energy sources, creation of economic conditions for obtaining balanced gains from supplying energy resources and power equipment to domestic and foreign markets, rational structure of import and export etc.

The problem of energy, economic and environmental safety is global and refers not only to each individual country but also to the world community as a whole. Its solution lies in the plane of mutually beneficial partnership of all participants: producing, transiting and consuming countries which have common organizational and management structures and take into account technological, economic and environmental risks of such cooperation. Thus, partner economies should not be overly specialized in the production or consumption of any definite type of equipment or energy source.

Such cooperation should be focused on the creation of infrastructure and management systems of joint cross-border markets which comprise the transportation networks of energy goods and services (gas and oil pipelines, power lines, trans-shipment terminals and so on), integrate domestic markets of different types of energy goods and services, radically reduce barriers on the way of energy goods and services flows between countries and regions, and improve the investment climate. This cooperation greatly enhances energy security, provides access to competitive energy sources and new technologies, and improves reliability of energy supply.

Additional organizational and material costs for the establishment and operation of cross-border markets should be paid off by lower costs of energy and equipment supply, while the arising risks of deeper differentiation between technologically developed countries and raw material exporting countries should be mitigated by coordinative regulation of these markets.

Energy services constitute an obligatory part of cross-border markets, which is oriented towards solving complex problems of efficiency, quality and reliability of energy supply and energy use improvement, with a view to coordinating interests of producers, suppliers and consumers of energy goods and services in energy sector and related sectors and sub-sectors of national economies. If we consider the sphere of energy services from this point of view, it becomes clear that the issues which are closely related to technical and technological components of identifying the reasons of inefficiency of cross-border markets functioning and focused on development of methods and measures for their prevention and/or elimination are only part of energy services which, if treated systemically, should include economic and environmental aspects of cross-border cooperation.

It is also clear that the international (trans-boundary, cross-border, etc.) cooperation provides new possibilities and creates new challenges in the area of energy services, which in today's conditions of energy market globalization have to be dealt with through interaction (cooperation) of competing energy service companies, taking into account differences in levels of countries' economic development, exchange rates, existing customs barriers, etc.

Mining industry of any country in its primary mission is raw material producing industry, which is characterized by complex resource-and-energy intensive technologies of extraction and processing of mineral resources. It is a branch of economy that requires modern control and optimization systems of mining production, and relates to all the above mentioned benefits and risks regarding the creation and operation of joint cross-border markets in the sphere of energy services.

Energy intensity of the gross domestic product (GDP) is one of the most generally acceptable basic energy-economic indicators of a country's economy performance efficiency. The energy intensity is the ratio of primary energy consumption to meet the manufacturing and non-manufacturing needs of the country to its GDP. The changing dynamics of this index helps to monitor the country's economic development (Geyts 2003).

Energy component (numerator) of the energy intensity index is a physical quantity accurately recorded by statistics, while the GDP (denominator) is calculated on the basis of the total value of goods and services produced by the country's economy during the year. The results of GDP calculations are presented in current, i.e. actual, prices (nominal GDP), and in prices comparable to those of the selected base year (real GDP). Their values for several years are usually defined in constant prices of the base year by using a chain of index-deflators, calculated as a ratio of GDP in current prices to GDP in constant prices.

Difficulties in comparison of the countries interaction efficiency on the cross-border markets arise due to the fact that GDP of each country is expressed in the national currency and should be converted before comparison into some international currency. The official exchange rates cannot be used for such conversion because in such case a number of factors related to the internal turnover of goods and services, capital flows, intrusions and interventions to the currency exchange market are neglected. Also, if we use official exchange rates, the numerous indicators that are defined in connection with GDP will be distorted. For example, in comparison of energy intensity indexes, poor countries will seem exceedingly wasteful, and this will lead to a biased forecast about future increase in energy consumption.

It should be added that using energy intensity index for national economies comparison is complicated not only due to exchange rates and technical efficiency of energy use, but also to the structure of industrial production, the level of common wealth, development level of the transport system, the country's geographical location, climatic conditions, the status of the shadow economy, etc. The situation is aggravated by the instability of exchange rates and disparity between prices/tariffs of goods and services and real costs for their production which is particularly noticeable for the transitional economies. In general, it is necessary to use the systemic approach to the valid comparative analysis and development of effective measures for improving energy efficiency of the national economy.

In international practice, the comparison of energy intensity of GDP in different countries is done, as a rule, with the help of purchasing power parity (PPP), whose value is calculated for many countries and regularly published by a number of international organizations, such as the United Nations, the World Bank and International Energy Agency (IEA) in terms of artificial monetary units - international dollars (Froot 1994, Stapel 2006). Herewith, the national exchange rate at PPP is determined by the amount of national currency that is spent in the country for the purchase of the same basket of consumer goods and services equal in quality to that bought by a resident of the United States for 1 USD.

For example, changes in production and consumption of primary energy in Ukraine during 1992-2011 are calculated and presented in Fig.1 in million (mln) tons of oil equivalent (toe), GDP in constant prices of 2005 - in billions (bn.) of Hryvnias (UAH) and billions of USD 2005 (PPP), and Ukraine's energy intensity of GDP - in kg of oil equivalent (koe) per USD 2005 (PPP).

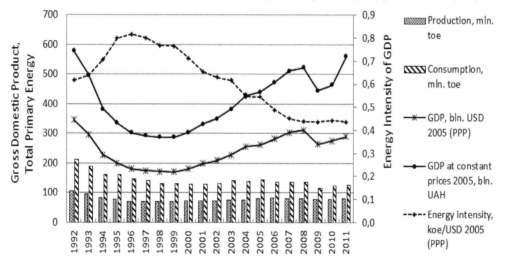

Figure 1. Total primary energy production and supply, GDP, energy intensity of GDP
Data source: www.yearbook.enerdata.net, www.ukrstat.gov.ua, authors' calculations

It can be seen that since 1996, energy intensity of GDP in Ukraine has been steadily improving (decreasing), but a deeper analysis of the economy shows that it is largely achieved not by increasing energy efficiency, but by attempts to ensure competitiveness of domestic goods and services by reducing wages expenditures and enterprises' circulating capital, freezing funds for modernization of production, using other negative means and instruments.

It is possible to investigate the energy-economic efficiency of national economies more comprehensively taking into account the environmental component of GDP. The most common internationally accepted environmental indicator (index) is the carbon intensity of GDP, which is defined by the ratio of the total emissions from burning fossil fuels (coal, oil and gas, etc.) in CO_2 equivalent to the GDP.

Carbon intensities of real GDPs at PPPs for Ukraine, Poland as its nearest neighbor, and average parameters for EU countries are shown in Fig. 2, from which it is obvious that Ukraine's indicator is far behind Europe.

Fig. 3 presents the results of comparative analysis of GDP's energy-economic effectiveness for several countries, which are potential (and actual, too) participants of cross-border cooperation with Ukraine at energy services markets, as reflected in the Cartesian coordinate system relatively to productivity of living labor needed to produce GDP in these countries. The values of energy-economic efficiency of GDP in 2009 for each country measured in thousands of USD 2000 at PPP per 1 toe of consumed energy are presented on the horizontal axis in Fig. 3, the vertical axis showing the productivity of GDP's living labor in 2009 in these countries, measured in thousands of USD 2000 at PPP per capita (GDP / population of the country). It can be seen that countries with developed economies and advanced technologies are placed at well ahead of Ukraine in the upper right corner of this coordinate system.

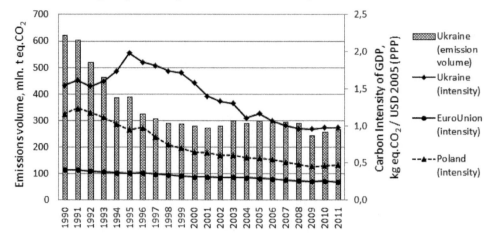

Figure 2. Carbon emissions and carbon intensity of GDP in Ukraine compared to some other countries
Data source: www.yearbook.enerdata.net

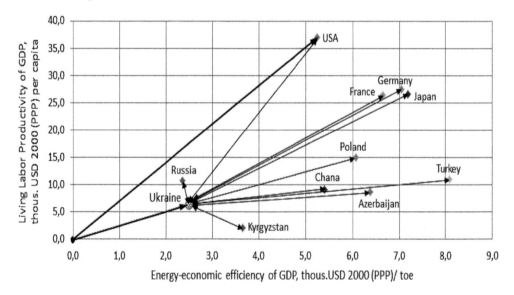

Figure 3. Efficiency and productivity of the GDP of some countries
Data source: www.iea.org

Energy Service Companies (ESCOs) are most effective and most widespread organizational forms of improving energy-economic efficiency of national economies, with established capacity to reach feasible results both in developed and developing countries. The most modern conceptual provisions of ESCO model were suggested by Shirley Hansen and Paolo Bertoldi, the ideologist of its distribution in Europe, who co-published more than a dozen books on the subject (Hansen 2009; Sivaev 2011).

ESCOs are commercial organizations (companies, enterprises), acting on the basis of the energy-service contract and providing a wide range of complex energy services that cover technical, economic, financial and legal aspects of designing, engineering, installation, commissioning, monitoring and verification of the results achieved from the implementation of innovative projects in the area of energy savings and energy infrastructure development at the industrial, communal and social facilities. ESCOs operate on the principles of energy management, performance-contracting, outsourcing, project financing, taking into account transaction costs and risk management, etc., using their own and/or attracting external sources and resources including financial, legal, material, technical, energy and labor ones.

Energy performance-contracting is the main form of providing complex "turnkey" services by ESCO, which offers their clients a full set of business projects to improve energy efficiency and energy conservation, the results of which are monitored (measured and verified) during the term of the contract, which guarantees that the savings achieved as a result of the projects implementation will be sufficient to cover the projects total costs (Hansen 2009; Sivaev 2011; Bachmann 2004). Manufacturers of energy-efficient equipment, energy supply companies, energy engineering and energy maintenance companies often operate as effective energy service companies. However, the energy services of ESCOs are fundamentally different from other types of energy service companies – suppliers.

Most of the problems related to the specifics of ESCOs activities at the domestic markets of energy resources, energy-efficient equipment and energy services are already solved in theory and practice as a result of ESCO-approach implementation. These issues are analyzed in publications of P. Bertoldi, J. Bleyl-Androschin, C. Bullock, G. Caraghiaur, J. Ellis, M. Evans, C. Goldman, S. Hansen, N. Hopper, B. Knox, M. Kolarik, P. Langlois, M. Lee, V. Lew, M. Magee, A. Marino, C. Murakoshi,H. Nakagami, N. Okay, J. Osborn, J. Painuly, H. Park, S.Rezessy, D.Schinnerl, S. Sorrell,A. Thumann, W. Turner, E. Vine, H. Zhao, and others. As for Ukrainian and Post-Soviet researchers, one should mention works by I.A. Bashmakov, S.V. Golikova,V.A. Zhovtyanskiy, M.P. Kovalko, E.E. Nikitin, A.V. Novoseltsev, A.V. Prakhovnik, S.B.Sivaev, V.A. Stepanenko, O.M. Sukhodolia, Y.I. Shulga,

describing the conceptual foundation and basic principles of ESCOs performance.

Regarding peculiarities of ESCOs' expenditures, the major role of transaction costs should be noted in the first place. They may be both internal and external, resulting from the company interactions in the energy services market, and include all necessary costs that are not directly related to the production of goods and services (production costs), but provide successful implementation of cross-border cooperation. The concept of transaction costs was introduced into economics by R. Coase in the 30-s of the previous century to substantiate the existence of firms and companies, which in their essence are hierarchically-structured systems with the administrative-command management. Among many fundamental publications on the theory of transaction costs we should first mention the works by K. Arrow, U. Asan, P.J. Buckley, M. Chapman,S. Cheung, R.H. Coase, N. Foss, C. Kadaifci, A. Kutlu, D.C. North, J.Ju, S. Sorrell, E. Wang, O.E. Williamson.

Despite the fundamental achievements in the theory and practice in the abovementioned areas of research, the problems of quantitative assessment of the effectiveness and efficiency of cross-border cooperation among ESCOs on domestic and international markets, evaluation and realization of comparative advantage potential of such cooperation still remain unsolved.

The purpose of this research is to justify the selection of energy-economic models of comparative advantage and performance-contracting in approaching the task of improving effectiveness of ESCOs cross-border cooperation, formalizing basic principles of building a system model on this basis, computing and presenting calculation results, confirming the efficiency of the proposed mode.

Applying Sorrell's methodological approach to constructing a performance-contracting model (Sorrell 2007), we can define generalized conditions of profitability (self-support) of ESCO approach in the following form: (1) the client's total cost savings, before and after signing the contract, should exceed the client's payments under the contract with ESCO (contract payments) in the amount of potential gain the client can receive under the contract; (2) the income of ESCO under the contract should be more than the total amount of expenses, incurred by ESCO, by the amount of its gain; (3) the client and ESCO overall production cost savings should exceed the overall increase in the transaction costs by the alternative costs of the client.

Formalizing the above conditions, we can write the generalized system of equations of ESCO interaction with the client as:

$$\sum_{k=1}^{n}\widehat{C}_{k}^{0} - \sum_{k=1}^{n}\breve{C}_{k}^{0} \geq \breve{P}^{0}; \quad \sum_{k=1}^{n}\widehat{C}_{k}^{0} - \sum_{k=1}^{n}\breve{C}_{k}^{1} \geq \breve{C}_{1}^{0} - \sum_{k=2}^{n}\breve{C}_{k}^{0}; \quad \sum_{k=1}^{n}\breve{C}_{k}^{1} \leq \breve{P}^{0} - \breve{B}^{1}, \tag{1}$$

where: the upper indexes 0 and 1 determine the client and ESCO; $\widehat{C}_{k}^{0}$, $\breve{C}_{k}^{0}$ – the client's costs before and after contract signing (this is indicated by superscripts ^ and ˇ), the lower index $k = \overline{1, n}$ lists the variety of cost types, and n– their quantity; $\breve{C}_{k}^{1}$ – corresponding ESCO costs; $\breve{P}^{0}$ – the client's contract payments (ESCO income); $\breve{B}^{1}$ – ESCO's bonus according to the contract; $k = (pr, tr, ds,...)$ is a sequence of indexes; pr – production costs; tr – transaction costs; ds – distribution costs.

In case of a client's cooperation with several ESCOs or an ESCO with several clients, for each case there are equation systems similar to (1), being in fact some kind of economic balance equations. It is clear that to complete mathematical description of the interactions between the client and ESCO, these equations have to be supplemented by the relevant equations of energy balances which restrict the acceptable modes of energy equipment operation and adjust the rules of mutual payments between the client and ESCO, caused mainly by changes in production amount and the corresponding decrease in the consumption of fuel and energy resources and subsequent lower loading of the energy equipment. These issues are specified as the objects of ESCO services contract.

While choosing the economic model of comparative advantage we should notice that ESCO by its main function is a service-oriented, and not a production company, so the bulk of its expenses are expenditures on intellectual work, while other types of costs are usually born by manufacturers of energy-efficient equipment and suppliers of fuel and energy resources. Therefore, the use of the Ricardian model (Krugman 2003), where the only factor of production is labor and which assumes that each technology can be characterized only by one coefficient of labor productivity – labor intensity, in our case, is justified and allows conducting a quantitative evaluation of the effectiveness and efficiency of cross-border cooperation between ESCOs.

Let us consider the peculiarities of the Ricardian model application, taking as an example two ESCOs, located in different countries, which intend to cooperate with each other, in buying and selling goods and services. We will define each ESCO by top index j (in this case j=1,2), supposing that each of the ESCOs, using existing technology, is able to provide two types of services, which can be defined by the lower index i=1,2. Then, the production of each of these services according to the Ricardian model will be characterized by its own factor of labor productivity τ_{i}^{j}, and the existing limits of production opportunities and options for the use of labor resources can be put in the form of :

$$\sum_{i}(\tau_{i}^{1} \cdot Q_{i}^{1}) \leq L^{1}; \quad \sum_{i}(\tau_{i}^{2} \cdot Q_{i}^{2}) \leq L^{2}, \tag{2}$$

where: Q_{i} – the amount of the service production during the working hours which are considered; L^{j}, $j = \overline{1, 2}$ – limited amount of labor resources, which the ESCO possesses.

In conditions of constant labor costs, as well as the amount of available labor resources, production opportunity curves are straight lines, whose equations assume the following form in coordinate axes $Q_{1}Q_{2}$:

$$Q_{2}^{1} = \frac{L^{1}}{\tau_{2}^{1}} - \frac{\tau_{1}^{1}}{\tau_{2}^{1}} \cdot Q_{1}^{1}; \quad Q_{2}^{2} = \frac{L^{2}}{\tau_{2}^{2}} - \frac{\tau_{1}^{2}}{\tau_{2}^{2}} \cdot Q_{1}^{2}. \tag{3}$$

Equations (3) showing that alternative costs of the first service production, expressed in units of the second one, or costs of substitution (substitution costs) of the first service by another one, are defined by the ratio of labor costs of the first and second services production $(\tau_{1}^{j} / \tau_{2}^{j})$ for either company. The amount of losses incurred by the company because of refusal from production of one service due to increased production of another service by 1 unit defines the alternative costs of utilizing available supplies. Adding these costs to the amount of the actual (explicit) production costs recorded in accounting books allows to specify the amount of the lost profit related to irrational distribution of available resources, refusal from other possibilities of their use in production of alternative services, etc.

To select from a number of possible combinations of the particular kind and amount of each of the two services feasible for ESCO to produce, the Ricardian model utilizes the comparison criterion of the relative price of one service measured per unit of the other service (the cost of the substitution of one service by the other). Let p_1^j and p_2^j be the prices of the first and the second services of the ESCO, then labor costs per hour λ_i^j will be equal to the costs of the first and the second services provided per hour: $\lambda_i^j = (p_i^j / \tau_i^j)$. If $(p_1^j / p_2^j) > (\tau_1^j / \tau_2^j)$, i.e. the costs of substituting the first service by the second one are more than alternative production costs of these services, then labor costs per hour will be higher in the production of the first service, and under the condition of $(p_1^j / p_2^j) < (\tau_1^j / \tau_2^j)$ – the same but for the second service. Taking into account the natural need of employees to receive higher payment for their labor, ESCOs according to the Ricardian model will specialize in production of one of the services and only under condition of equal payment for both services.

Let us assume that:

$$(\tau_1^2 / \tau_2^2) > (\tau_1^1 / \tau_2^1), \text{ or the same as } (\tau_2^1 / \tau_2^2) > (\tau_1^1 / \tau_1^2). \tag{4}$$

The first inequality of two ratios in (4) is called the formula of relative advantage, which shows that the labor costs for the substitution of one unit of the first service by the second service in ESCO1 are lower than in ESCO2. In this case, ESCO1 turns out to be more effective in production of the first service than the second one (the second inequality), i.e. has comparative advantage over ESCO2 in production of this service.

Thus, to select a more beneficial company for production of the first service in conditions of cross-border cooperation is not enough just to compare labor costs of both ESCOs necessary for production of one unit of this service and to make a decision on starting its production on the basis of such absolute advantage. According to Ricardo's theory, company should rely on the principle of comparative advantage, taking into account production costs of each of the services that both companies are potentially able to produce (provide to the market).

Under the principle of absolute advantage, prices for these services depended solely on the internal factors of production, while in conditions of cross-border cooperation prices are established in terms of market supply and demand factors for the services that now are exported or imported. e In market conditions, the ESCO with lower prices begins to compete with other companies for the cross-border market share, which in turn stimulates the relative price equalization (equilibrium) for these services. The final result of equilibrium is that market prices are somewhere between the existing relative prices, and their definite level will be conditioned by the amount of mutual demand and supply. That is, the price of imported services will depend on the price of the service that a company wants to export, to pay for the import, and the ratio of these prices will depend on domestic demand for these services within the sphere of influence of each ESCO.

The following formulas may be used to calculate cross-border cooperation values:

$$\breve{Q}_1^1 = (1 + \delta_1^1) \cdot Q_1^1; \, p_{2/1}^1 = ((p_1^2 / p_2^2) \cdot p_2^1 - p_1^1) \cdot d_{2/1}^1; \, \Delta_1^1 = \delta_1^1 \cdot Q_1^1; R_{2/1}^1 = p_{2/1}^1 \cdot \Delta_1^1, \tag{5}$$

where: $\breve{Q}_1^1$– amount of service1 extended production by ESCO1; $\delta_1^1 = (\tau_2^1 / \tau_1^1 - \tau_2^2 / \tau_1^2)$ – coefficient of service1 extended production; $p_{2/1}^1$– price of cross-border substitution of service2 by service 1; Δ_1^1– amount of alternative service1 production; $R_{2/1}^1$ – amount of alternative revenue from the substitution of service2 by service1;

$$d_{2/1}^1 = \exp(-\tau_1^1 \cdot (\breve{Q}_1^1 - Q_1^1) / (\tau_1^1 \cdot Q_1^1 + \tau_2^1 \cdot Q_2^1))$$

– equation of cross-border demand graph.

On the basis of equations and formulas (1) – (5) the computer model of comparative advantage of ESCO cross-border cooperation has been realized in tabulated Excel-processor with the use of Solver optimizing tool, which allows to make series of parametric calculations and presents the results of calculations in tabular and graphic forms. For example, the results of the model calculations of the amount of alternative service1 production (a/service 1) as well as limits of ESCO1 production possibilities before and after cooperation with ESCO2 are presented in Fig.4.

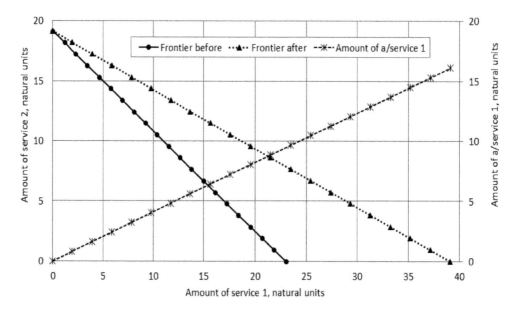

Figure 4. ESCO1 growth in amount of service 1 production (provision)
Data source: authors'research

Fig. 5 presents the calculated amount of alternative revenue $R^1_{2/1}$ (a/revenue 2/1), which is measured by the right hand vertical axis in foreign currency 1 (the currency of ESCO1), from the price of cross-border substitution $p^1_{2/1}$ (t/substitution price 2/1 in currency1), calculated by the exponential curve $d^1_{2/1}$ of cross-border demand (t/demand 2/1) and from the amount Δ^1_1 of alternative service 1 production (a/service 1), necessary for cross-border substitution, all that being the result of change in the amount $\bar{Q}^1_1$ of service1 production by ESCO1.

It is obvious that in the considered range of extended production (provision) of service1 the amount of alternative revenue of ESCO1 from its cross-border cooperation with ESCO2 is steadily growing to constitute 28.0 monetary units in currency1 for 39.1 natural units of service1 production at the price of cross-border substitution of 1.74 monetary units in currency1. At the same time, the amount of alternative revenue of ESCO2 from its cross-border cooperation with ESCO1 is 15.45 monetary units in currency 2, which is received as a result of production (provision) of 73.7 natural units of service 2 at the price of cross-border substitution of service1 by service2 in the amount of 0.39 monetary units in currency 2.

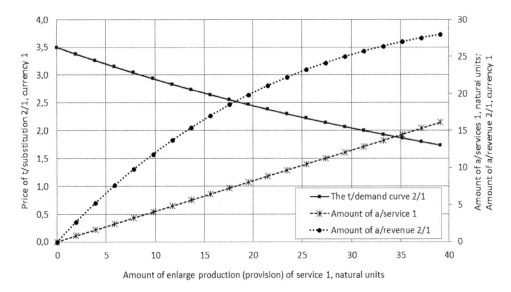

Figure 5. ESCO1 relationships between the alternative revenue amount and the price of cross-border substitution as well as the amount of alternative service1 production (provision)
Data source: authors' research

Thus, the improved energy-economic model of ESCOs cross-border cooperation has been developed. The model is based on a systemic conjunction of theoretical statements of performance contracting, which is one of powerful tools to improve energy efficiency of industrial private and public enterprises. Applying provisions and principles of economic comparative advantage allows ESCOs to expand the spheres of their activities and to increase energy-efficient goods and services production. As a result, ESCO approach enables not only to find new opportunities to improve energy efficiency and energy conservation, but also to develop a new, previously not available, mechanism of obtaining economic and environmental gains from cooperation between ESCOs, which is an important factor in increasing the energy and economic safety. It is important that the model of ESCOs cross-border cooperation can be applied to definite areas, regions, and other administrative-territorial divisions and units, in one or several countries.

The prospects for further research in the field of energy services are primarily conditioned by the need to supplement the model with equations of energy balance in order to use this extended model in technical and economic calculations of benefits from attracting investments into upgrading equipment in mining industry, taking into account comparative advantages that arise as a result of using the saved fuel and energy resources as a substitution product.

REFERENCES

Geyts, V. M. 2003. *Economy of Ukraine: Strategies and policies for long-term development* (in Ukrainian). Phoenix: Institute of Economics and Forecasting: 1008.

Froot, K. & Rogoff, K. 1994. *Perspectives on PPP and Long-Run Real Exchange Rates.* Cambridge: National Bureau of Economic Research: 55.

Stapel, S. & Roberts, D. 2006. *Eurostat-OECD Methodological Manual on Purchasing Power Parity.* Paris: OECD PUBLICATIONS: 280.

Hansen, S., Bertoldi, P. & Langlois, P. 2009. *ESCOs Around the World: Lessons Learned in 49 Countries.* Lilburn: The Fairmont Press: 337.

Sivaev, S.B. 2011. *The establishment and activities of energy service companies and performance contracts in Russia* (in Russian). Moscow: WWF, Vol. 1: 111.

Bachmann, J. & Novoseltsev, A. 2004. *Partnership and Incentives: Making Performance Contracting Work in Ukraine* (in Ukrainian). Energy Engineering, Vol.101, Issue 6: 49-70.

Sorrell, S. 2007. *The Economics of Energy Service Contracts*. London: Energy Policy, Vol.35, Issue 10: 507-521.

Krugman, P. & Obstfeld, M. 2003. *International Economics*. Boston: Pearson Education: 782.

Energy Efficiency Improvement of Geotechnical Systems – Pivnyak, Beshta & Alekseyev (eds)
© 2013 Taylor & Francis Group, London, ISBN 978-1-138-00126-8

The electrochemical cathodic protection stations of underground metal pipelines in uncoordinated operation mode

A. Aziukovskyi
State Higher Educational Institution "National Mining University", Dnipropetrovs'k, Ukraine

ABSTRACT: The paper analyses the joint work of cathodic corrosion protection stations of underground metal pipes. It is shown that the risk of corrosion on the border line of the two stations which operate in uncoordinated modes increases.

1 INTRODUCTION

The problem of gas supply system pipelines in general is formulated as the extension of operation of underground pipelines with minimizing of uncontrolled losses of substances transported. The natural gas pipeline system in Ukraine is the oldest in Europe - more than 40 percent of pipelines have been in operation for more than 12 years, 29% - for 13-22 years, 14.8% - 23-32 years, 15.2% - 33-47 years, 0,1 % - for more than 50 years. General metal loss from corrosion in industrialized countries can be compared with the metal consumption in the development of the metal industry. However, it does not take into account the loss of resources, decreased productivity and product quality as a result of industrial accidents and environmental damage. That is, considering indirect costs, corrosion is taking up to more than a tenth of national income in developed countries. In Ukraine, the problem of protection of underground metal constructions from corrosion acquires critical dimensions against the background of changes in the market of energy.

2 FORMULATING THE PROBLEM

Analysis of the recent research and publications indicates that there are still technical problems related to the area of research that are still unresolved. The development of modern electronic components leads to the introduction of the controlled converter in the systems of cathodic protection from corrosion. Certain publication (Jala 1998; Peydzhak 2000) mention the possibility of the corrosion of underground metal pipelines reduction due to uncoordinated elements of active cathodic protection system.

Isolation unsolved aspects earlier of the total problem is incomplete as attention to the fact, that quite often in one system of cathodic corrosion protection integrated cathodic protection stations of various types nobody did not pay. Therefore, the task is to evaluate the possibility of fluctuations in the values of polarization potential for underground metal piping that is under the influence of cathodic protection stations of various types.

3 MATERIALS UNDER ANALYSIS

Underground electrochemical corrosion is the destruction of metal being to its interaction with the corrosive environment. Thus, speed of ionization of metal components and restore anion of oxidative corrosion protection depends on the electrode potential. Electrochemical corrosion is accompanied by the electrical current. For underground structures laid in the soil, characterized by the following types of electrochemical corrosion:

– underground corrosion – is the corrosion in the soil caused by electrochemical micro-and macro-pairs that emerge on the metal in the of its contact with corrosive environment that acts as electrolyte. Corrosive pairs result from heterogeneity of metal texture, soil structure or composition of the electrolyte, the differences in temperature, humidity and air permeability of the soil along the route of the pipeline.

– underground biocorrosion caused by microorganisms that affect the metal. Usually the process is completed by of electrochemical corrosion.

– electrical corrosion, of the processes of corrosion in metal underground facilities caused by leakage currents of electrified rail transport or other industrial electric equipment, existing two forms:

    o   corrosion influenced by ground current;

    o   corrosion due to external current which occurs when the current coming flows through the metal and electrolyte.

Depending on other conditions and the type of corrosion, can distinguish continuous, local uniform, uneven and contact corrosion. Interaction of soil electrolyte with metal in any form of the electrochemical corrosion can be divided into two processes: anodic and cathodic (Tsikerman et al 1958; Strizhevskij et al 1967). Anodic process consists in anode metal, in transition the electrolyte in the form of cations with equivalent number of electrons in the metal. Free electrons move along the metal from the anode to the cathode areas, and are involved in the reduction reaction. Cathodic process is accompanied by the absorption of excess electrons by depolarizers of electrolyte (atoms, molecules or ions that can be restored on the cathode). Mainly in the underground corrosion, depolarization occurs due to the discharge of hydrogen ions (hydrogen depolarization) and ionization of oxygen atoms (oxygen depolarization). Because soil in most part has a neutral reaction, the process of corrosion of underground metal constructions is often accompanied by oxygen depolarization. Electric current flow between the anodic and cathodic areas is caused by the movement of electrons in the metal from the anode to the cathode, and in the electrolyte - by the ions movement. Anodic and cathodic processes in most cases occur at different sites. Surface of the metal consists of a number of micro-and macro-pairs and the corrosion rate depends on the number of such pairs and intensity of their work. This mechanism of the electrochemical corrosion is called heterogeneous process. In the homogeneous electrochemical mechanism of corrosion, the anodic and cathodic processes occur in the same area. It is important that whatever the mechanism of corrosion the material effect is manifested only on the anode. On the cathodic areas, where the process of depolarization taking balance metal losses are not observed. Pure metals are normally oxidized by homogeneous mechanism. Impurities and micro-heterogeneity of steel, and soil, heterogeneity of oxide layers on steel cause emergence of oxidation areas. Different composition of the electrolyte and different aeration of individual areas on the contact surface of dissimilar metals create conditions for occurrence of corrosion. Explicit anodic and cathodic areas are observed in the presence of ground currents. The corrosion process should be assessed from the viewpoint of heterogeneous electrochemical corrosion mechanism.

During the process of underground corrosion, metals are usually present are not in solution of their salts, but in solutions of other electrolytes. The magnitude of the potential is mainly influenced by the concentration of hydrogen ions (pH), and various other processes (hydrogen emission, reactions leading to the emergence of films). In such cases, the value of sustainable equilibrium potential will be different from the normal one. Stationary (or natural) potential - is the equilibrium potential of the metal in the given concrete electrolyte in the absence of external current. In this case, potential current that causes anodic dissolution of metal areas, is fully compensated by the current, which is needed to restore oxygen in the cathodic areas. Most often, stationary potentials are measured relative to copper electrode sulfate, which does not change its potential due to the current passing through it, and has a certain equilibrium potential, which is not polarized (+0.3 V. relative to normal hydrogen electrode). The values of fixed potential and low-carbon steels in soil in relation to the electrode are approximately the same, and in many cases close to the value of - 0.55 V.

Depending on the of insulating coatings, composition and soil moisture, this value can vary by ± 0,2 V. The shift of potential in to negative direction occurs with increasing heterogeneity of cover and humidity, decreasing of specific resistivity and air permeability of the soil. Thus, in moist soils, stationary potential is more negative than in the sandy soil. In the first case, its value can reach - 0.7 ... 0.8 V, in the second case - 0.3 ... 0.4 V. The deposition of corrosion products on the metal surface helps shift potential into positive values. The more negative is the potential of the metal structures in the ground, the more it is prone to oxidation. However, the value of stationary steel potential in soil in the absence of ground currents or currents of cathodic protection is not an unambiguous indicator of corrosion hazard for the underground pipeline. The potential of any electrode, which was placed in the electrolyte changes when an electric current passes through. Potential reaches the value called "polarization potential" or "under-current" potential, unlike the equilibrium electrode potential with no current. In this case, the electrode gets polarized.

Polarization is any change on the metal surface, which is located in the electrolyte. Polarization is the result of input or output current and is related to changes in metal-electrolyte potential. Slowing down the any stage of the corrosion process is the cause of polarization. The most common is concentration polarization, which is related to the concentration of metal ions, hydrogen or oxygen in the space around the electrode. In the anodic polarization anode potential shifts in a positive way, during cathodic polarization - cathodic potential shifts in the negative direction. Polarization accompanies electrochemical corrosion any type, regarding the destruction of metal, reducing the rate of electrochemical processes. Electrodic process that reduces the value of polarization is a process of depolarization. Depolarization reduces the shift of electrode potentials and increases the rate of corrosion. The protection of underground pipelines from corrosion can be divided into passive and active. The passive corrosion protection is pipe insulation from contact with the surrounding soil in order to limit the influence of ground currents. Active protection creates a protective potential of underground metal pipe in relation to the environment.

Isolation of the underground metal pipe from contact with the environment is achieved by selecting a protective coating that meets the conditions of operation. Rational choice underground metal pipelines location and the use of special laying methods: in canals, blocks, tunnels, sewers, etc. Laying of underground piping should be along done on the routes with the least danger of corrosion. It means avoiding areas with highly aggressive soils, dumps and areas industrial waste and utility companies runoff.

The state of metal structures corrosions is estimated by: visual inspection, metallographic methods, chemical, electrochemical methods, mechanical tests along the following criteria:

- local index of corrosion rate - the number of corrosion zones per unit of metal surface for a preset period of time;
- the indicator of the corrosion depth - the value which is characteristic of the average or maximum depth of corrosions area;
- metal proneness to corrosion - lifetime till the beginning of the corrosions (if the zone of corrosion reaches more than 1% of the total surface area of the metal structure);

- change in metal mass – due to losses from corrosion processes;
- current rate of corrosion - current per surface area unit of the metal structures.

In order to protect from corrosion, several methods can be used:

- Methods that affect the environment and operating conditions;
- Methods that affect metal;
- Combined methods.

Metal coating is most widely is used. Its disadvantages include short life of the coating, the need to control its state (which is quite difficult to implement, for distributed metal pipeworks), susceptibility to mechanical damage. The alternatives for corrosion protection of underground pipeline systems are the methods that affect the environment. The branching system of underground pipelines is one of the main factors causing additional difficulties in the process of corrosion protection in terms of electrochemical theory.

Cathodic protection can be implemented using DC sources external to the pipeline, which is thus protected from galvanic corrosion, or using galvanic electrodes (anodes) which act as the electrochemical constant current sources. Both methods are based on the common principle but have different field of application and methodology. External electric current source provides better cathodic protection; it is used during intense anode situation observed over large areas. The specified type of protection from galvanic corrosion is used in conditions that are caused by ground currents. The protection by anodic grounding is applied in situations for which significantly lower rates of capacity are typically used to control soil corrosion in areas with small anodes zones. The cathodic protection of the metallic pipelines is ensured by the following elements (Fig.1):

- Station Cathodic Protection (SCP - DC source, measuring devices, electric apparatus);
- The earthing device (anode);
- Connecting lines (between the line of electric supply, rectifier, anode earthing switch and metal structures).

In calculations of electrical parameters of the underground metal pipe specific conductivity of the earth is referred to as: σ. Since the soil is not homogeneous in composition along the pipeline, the average conductivity is used in the calculations computing electrical parameters of the pipeline, we distinguished between the primary and secondary parameters. The former include the geometrical dimensions of the pipeline, its route options and the conductivity of the pipe material.

Permeability of the pipe material and frequency of the electromagnetic field are also included in computations. Specific conductivity is calculated as (Strizhevskij et al 1967):

$$\sigma = \frac{10^6}{\rho} \qquad (1)$$

where ρ - specific conductivity.

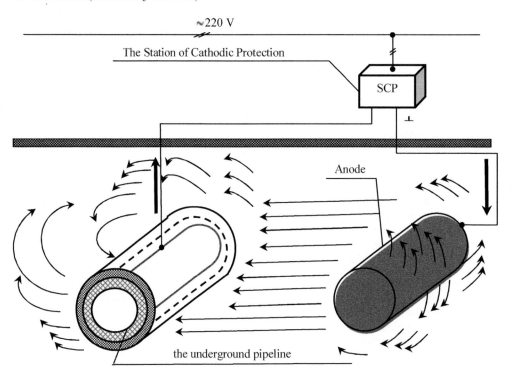

Figure 1. The system of protection from electrochemical corrosion

If the pipeline contains curved sections, their radius must be taken into account as r_t - the outer radius of the underground pipeline. Furthermore the thickness of the walls of the underground pipeline (h_T) and the depth of piping in the ground (H) must also be considered. The important parameter of insulation resistance of underground pipeline is determined by the following formula:

$$R_i^T = \frac{h_i}{2\pi r_T \sigma_i^T} \qquad (2)$$

where h_i - total layer thickness of piping insulation; σ_i^T - conductivity of the insulation, which is possible only if the insulation layer is uniform along the length of the pipeline. However, in practice, these conditions are not met. Insulation resistance of the pipeline, depending on the type of insulation coating and its condition varies from 10 to 10000 $[Ohm \cdot m]$.

Secondary parameters are determined by the total resistance of the pipeline Z_T, by the resistance of the transition between the underground pipeline and the point in the ground $R_{T.Z}$. The values of the propagation constant α_T and the resistance pipeline (along its length) R_p^T are assumed to be known. Underground pipeline can be considered a hollow metal cylinder. The impedance of such cylinder depends on the amount of absorption in the conductor:

$$\gamma_T = \sqrt{\frac{\omega\mu\sigma_T}{2}} \qquad (3)$$

where $\sigma_T = 7.7 - 10^6 \ 1\!\!/\!\!_{ohm \cdot m}$ - for materials that are used in the manufacture of pipes. The conductivity of the pipeline: $\rho_T = 0.13 \ ohm \cdot mm^2\!\!/\!\!_m$.

The wall thickness of the pipe may be accepted as $h_T = 0.004 \, m$. The impedance of a cylindrical conductor is the sum of:

$$Z_T = Z_{p.T} + \tilde{Z}_T \qquad (4)$$

where $Z_{p.T}$ - the surface resistance of the pipe, which is calculated by:

$$Z_{p.T} = \frac{(1-i)\gamma_T}{2\pi r_i \sigma_T} ctg\left[(1-i)\gamma_T h_T\right] \qquad (5).$$

$\tilde{Z}_T$ - current conductive impedance, which is influenced by the external environment, and calculated as: $\tilde{Z}_T = \frac{\omega\mu}{8} + i\frac{\omega\mu}{2\pi} \ln\frac{1.3}{\gamma r_T}$ where

$\mu = 4\pi \cdot 10^{-7}$ - the permeability of the ground;

$\gamma = \sqrt{\frac{\omega\mu\sigma}{2}}$ - attenuation coefficient in the ground; σ - conductivity of the soil.

Considering that the resistance of the pipe per length unit for DC is defined as:

$$R_T^0 = \frac{1}{2\pi h_T r_T \sigma_T} \qquad (6)$$

the impedance of the conductive hollow cylinder will be calculated by:

$$Z_T = R_T^0 \Theta\left[(1-i)\gamma_T h_T\right] + \frac{\omega\mu}{8} + i\frac{\omega\mu}{2\pi}\ln\frac{1.3}{\gamma r_T} \qquad (7)$$

where Θ - a complex function of the dimensionless parameter $\gamma_T h_T$:

$$\Theta = \Theta'(\gamma_T h_T) + i\Theta''(\gamma_T h_T) = \\ = (1-i)\gamma_T h_T \times ctg(1-i)\gamma_T h_T \qquad (8)$$

If the cathodic protection against galvanic corrosion is operated on DC, the variables Θ' and Θ'' take the values 1 and 0 respectively. By

substituting in (7) values ω, μ, and using (8) we obtain the value of the total pipe resistance:

$$Z_T = R_T' + iR_r'' \qquad (9)$$

$$R_T' = R_T^0\Theta' + 4.9\cdot10^{-5} \qquad (10)$$

$$R_T'' = R_T^0\Theta'' + 6.3\cdot10^{-5}\ln\frac{93}{\sqrt{\sigma_T r_T}} \qquad (11)$$

where R_T^0 - loop resistance tubing, $ohm\!\!/\!\!_m$; σ - conductivity of the ground, $ohm \cdot m$; r_T - the pipe radius, m.

For DC systems of cathodic corrosion protection the following formula is applied:

$$Z_T = R_T^0 \qquad (12)$$

If we express the pipeline impedance through its components - active resistance of pipeline and its inductance, we get the formula:

$$Z_T = R_T + i\omega L_T \qquad (13)$$

Taking into account equations (13-15), we obtain expressions which allow to define the electrical parameters of the pipeline (Strizhevskij et al 1967):

$$R_T = R' = R_T^0\Theta' + 4.9\cdot10^{-5} \qquad (14)$$

$$L_T = \frac{1}{\omega}R'' = R_T^0\frac{\Theta''}{\omega} + \frac{6.3}{\omega}10^{-5}\ln\frac{93}{\sqrt{\sigma r_T}} \qquad (15)$$

where $\Theta' = \gamma_T h_T \dfrac{ch2\gamma_T h_T + \sin 2\gamma_T h_T}{ch2\gamma_T h_T - \cos 2\gamma_T h_T}$;

$\Theta'' = \gamma_T h_T \dfrac{ch2\gamma_T h_T - \sin 2\gamma_T h_T}{ch2\gamma_T h_T - \cos 2\gamma_T h_T}$.

Transient resistance of the pipeline electric current is defined as the sum of the insulation resistance and the resistance of the cross-section of the conductor:

$$R_P^T = R_i^T + R_L^T \qquad (16)$$

where $R_L^T = \dfrac{1}{\pi\sigma}\ln\dfrac{1.12}{\gamma\sqrt{r_T H}} \qquad (17)$

In (17) the value, which characterizes the depth of the pipe in the ground, is designated as H.

Attenuation coefficient in the ground is calculated as:

$$\gamma = \sqrt{\frac{\omega \mu \sigma}{2}} \qquad (18)$$

If the pipeline is located on the surface of the ground, the value H should be equal to r_T. Substituting the values $\gamma, \omega = 100\pi, \mu = 4\pi \cdot 10^{-7}$, after simplification we can write the formula:

$$R_P^T = R_i^T + \frac{1}{\pi\sigma} \ln \frac{2.8}{\sqrt{r_T H \sigma}} \qquad (19)$$

where the value R_P^T is measured in $ohm \cdot m$.

At the same time, knowing the transition resistance "pipe-to-earth" and resistance of the distribution, we get:

$$\alpha_T = \sqrt{\frac{Z_T}{R_P^T}} \qquad (20)$$

Given that:

$$Z_T = |Z_T| e^{i\varphi_T} \qquad (21)$$

where φ_T - this is the phase of the impedance of the underground pipeline impedance, which is calculated using the formula:

$$\varphi_T = arctg\left(\frac{R''}{R'}\right) \qquad (22)$$

The phase value of the line impedance can vary from 75^0 to 83^0. Given the most commonly used pipe diameter, the values of the phase vary between 81^0 to 83^0. We can express the average phase of the propagation constant as:

$$\frac{\widetilde{\varphi}_T}{2} = 41^0 \qquad (23)$$

Assuming the value of $\dfrac{\widetilde{\varphi}_T}{2} = 45$, which is less than the error of 8 percent, we obtain the simplified expression for calculation of the pipeline propagation constant:

$$\alpha_T = (1+i)\sqrt{\frac{|Z_T|}{2R_P^T}} \qquad (24)$$

Bearing in mind the constant of the potential distribution and current on the underground pipeline and the fact that as a rule cathodic protection stations are located close to each other, the pipeline can be presented in the form of system with parameters. Electrical equivalent circuit of the system "pipeline - the earth" in general can be presented in Fig. 2. Single-phase irreversible or reversible rectifiers (uncontrolled or semi- controlled) as rectifiers, that are part of the stations electrochemical protection is used. Usually, semiconductor elements of Station Cathodic Protection are connected in a bridge rectification circuit and fed from the secondary winding of the transformer with, the voltage:

$$U = \sqrt{2} \cdot U_2 \cdot \sin(\omega t) \qquad (25)$$

where U_2 - RMS voltage; ω - angular frequency of the supply voltage. Constant component of voltage which is rectified or its average value U_d, is equal to the integral of the voltage function, changes over time during the period 2π divided by his period:

$$U_d = \frac{1}{\pi} \cdot \sqrt{2} \cdot U_2 \cdot \sin(\vartheta) d\vartheta = \qquad (26)$$
$$= 2 \cdot \frac{\sqrt{2}}{\pi} \cdot U_2 = 0,9 \cdot U_2$$

Assuming equivalent circuit of underground metallic pipeline (Jala 1998; Peydzhak 2000) to be RLC loop (Fig. 2), let us consider the processes occurring in the system "single phase rectifier - load". The output voltage at the output of single-phase unguided rectifier has pulse shape (Fig. 3). When controlled rectifier is used, there appears some distortion of the rectified voltage (line «vs» in Fig. 3).

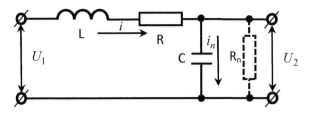

Figure 2. Equivalent circuit of the system "underground metal pipes - earth"

Thus, underground piping is protected from galvanic corrosion by forming protective capacity. The potential must be constant along the length of the underground pipeline that is protected. However, the potential value has to be constant in time and stay within the range of - 0.87 ... -2.5 V (Sklyarov 2002; Verbenets 2011; Strizhevskij et al 1967). It should be noted that the same system of underground pipelines can be protected by several stations with different type of rectifiers. In this case, one may different obtain levels currents that flow from the surface of the pipeline and the surrounding earth. For example, in the case when controlled rectifier used as a voltage source (Fig. 4), the average rectified voltage can be written as:

$$U_d = \frac{1}{\pi} \cdot \int_0^\pi e_2 d\omega t = \frac{1}{\pi} \cdot \int_0^\pi E_{2m} \sin(\omega t) d\omega t = \frac{E_{2m}}{\pi} \cdot \left[-\cos \pi + \cos \alpha \right] = \frac{E_{2m}}{\pi} \cdot \left(1 + \cos \alpha \right). \qquad (27)$$

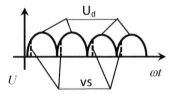

Figure 3. The rectified voltage

It is known (*Jala RM, 1998*) that fluctuations of protective potential values (in time and along the length of a pipeline) deteriorate the corrosion state of the object protected. Given the underground pipeline is simultaneously under potentials that vary in time by different laws, it is possible, that oscillations of potential can occur on the border of two adjacent stations of cathodic protection (Fig. 5).

For long pipelines, the SCP systems are applied to ensure high quality protection against galvanic corrosion. The protective potential of each SCP is changing along the length of the pipeline under the curve $\varphi(x)$ and should not be less than the minimum allowable value of φ_{min}. Otherwise it will lead to partial protection against galvanic corrosion of underground metallic pipeline. On the border line of the SCP impact, the oscillating processes which virtually do not influence on the quality of protection in the immediate vicinity of the SCP, can cause deviating values of the protective potential of less than φ_{min}.

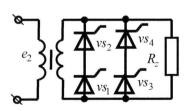

Figure 4. The controlled rectifier

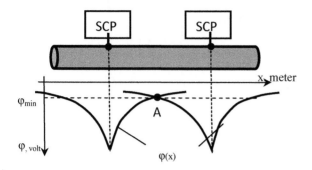

Figure 5. The underground pipeline protected by two cathodic protection stations

On the basis of mathematical expressions and formulations above, the operation of two cathodic protection stations was simulated. The simulation takes into account different voltage converters types that were used in these stations. The separate part of the underground pipeline of less than ten kilometers length is examined. The element of the underground pipeline is protected by the cathodic protection station with diode elements and thyristor cathodic protection station (Fig. 4). Figure 6 shows a plot the time variation of the underground pipeline potential.

s we see from the plot, the potential not only varies significantly but also changes its sign. This phenomenon indicates the presence of anodic areas on the pipeline which contribute to the corrosion danger. Modeling was performed at different opening angles of thyristors. Practical measurement of the potential variation in time is shown in Fig. 7, which demonstrates, that the shape of signals (they change over time) is the same. There is also a change of sign and

magnitude of the underground pipeline potential. Under the influence of the alternating signals, which is the result of uncoordinated operation of cathodic protection stations, the underground pipeline is exposed to significant threat of corrosion. The alternating component is responsible for the presence of the anode zone stray along the pipeline. All of the above mentioned increase factors the transfer of metal from the pipe into the ground which reduces the life of underground pipelines, necessitating their more frequent screening for the presence of corrosion damage. It should be taken into account that often rectifiers of voltage cathodic protection stations are connected to different phases. In this case, the rectified voltage of two stations are shifted because of the shift between phases in AC power line by 120 degrees. If it is required to ensure the protection of underground metal pipes of considerable length, it is necessary to remember that the maximum protective potential should not exceed - 1.72 V.

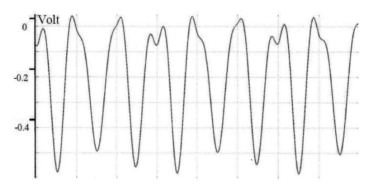

Figure 6. The plot of the time variation of the potential

Otherwise, this brings about overcompensation of the protected underground pipeline. If the potential of underground metal pipeline negative

to the point of measurement (copper sulfate) decreases significantly, the insulation begins to peel off. We should also separately consider a

54

system of underground pipes that are in the area of currents. Numerical characteristics of the corrosion danger is taken to from the surface density of leakage current from an underground pipeline. Processes be observed in a system of underground pipelines that are under the influence of variable stray currents are different from the processes observed in the system, acted upon by DC. According by appearance of the variable component of the curve in the potential an underground pipeline could mean that the real level of corrosion differs from the calculated one. The presence of a positive value in the curve, which describes the potential change the over time, indicates that for a certain period of time the underground pipeline remains unprotected against galvanic corrosion.

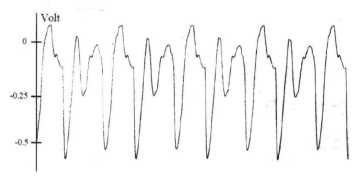

Figure 7. Change of the underground pipeline potential in time

4 CONCLUSIONS

Simultaneous operation of work stations for cathodic protection from electrochemical corrosion, their different circuit design solutions are defining fluctuations of the potential along on the length of underground pipeline. Thus further investigation of the ways of the aligning work modes of SCP for reducing variation of potential polarization, is essentially promising.

REFERENCES

Tsikerman, L., Nicholas, K. & Razumov, L. 1958. *Calculation of cathodic protection.* Ed. Prof. Mikhailova, M.I. Moscow: Mashinostroenie: 140.

Jala, R.M. 1998. *Current status and problems of corrosion control of underground pipelines.* Problems of corrosion and corrosion protection of structural materialstion: IV Internat. service ID whist "Corrosion '98". Lviv: FMI National Academy of Sciences of Ukraine: 411-414.

Peydzhak, J.T. 2000. *Uncoordinated operation mode of cathodic protection increases the likelihood of damage.* Oil and Gas Technologies, Issue 1: 93.

Sklyarov, S.A. 2002. *Mathematical models and information technology systems, automated control of corrosion protection of pipelines:* Dis. for the degree of Doctor of Science: 05.13.06 (in Ukrainian). Kharkov: 168.

Strizhevskij, I.V. & Dmitriev, V.I. 1967. *Theory and calculation. Influence of Electric Railway on underground metallic structures.* Moscow: Mechanical Engineering: 227.

Verbenets, B.J. 2011. *Non-contact method and apparatus for controlling corrosion protection of underground pipelines*: Dis. for the degree of Doctor of Philosophy: 05.11.13 (in Ukrainian). Lviv: 106 .

Energy Efficiency Improvement of Geotechnical Systems – Pivnyak, Beshta & Alekseyev (eds)
© 2013 Taylor & Francis Group, London, ISBN 978-1-138-00126-8

Enhancing of energy efficiency of utilizing waste heat in heat pumps

W. Czarnetzki
Esslingen University of Applied Science, Esslingen, Germany

Yu. Khatskevych
State Higher Educational Institution "National Mining University", Dnepropetrovs'k, Ukraine

ABSTRACT: Factors that influence on heat pump energy efficiency are analyzed, experimental data are given. The possibility to increase energy effectiveness of heat pumps due to operational control that takes into account parameters of heat source and consumer is shown. It is proposed to investigate modes of heat pump operation with the help of mathematical model. This model presents refrigerant as a two-phase medium and analyses changes in physical parameters along the evaporator and the condenser. It permits to explore modes with incomplete phase transformation of the refrigerating medium which occur in real processes.

1 INTRODUCTION

Utilization of waste heat in geotechnical, technical systems and technological processes is a way to improve energy efficiency. For geotechnical systems of mines the amount of waste heat is significant. At the same time energy for heating or hot water preparation is required. So one of the perspective tendencies is to use heat pump to utilize waste heat and to use it further in heating systems. Heat pump technology is widely used due to its energy efficiency and environmental safety. But when working conditions differ from rated values, efficiency of the heat pump decreases. In the most cases heat demand and amount of waste heat are unstable, which causes increase in specific energy consumption of the heat pump. It is proposed to analyze parameters that influence heat pump energy efficiency and offer approaches to improve it.

2 ANALYSIS OF PARAMETERS THAT INFLUENCE HEAT PUMP ENERGY EFFICIENCY

The main components of the heat pump affecting efficiency are: evaporator, compressor, condenser and throttling valve. Also it's important to take into account that heat pump works as a part of heating system. Parameters of the heat pump and heat consumer produce a substantial impact on energy efficiency.

Thus the main parameters influencing energy efficiency of the heat pump are:

1.   compressor efficiency;
2.   construction and characteristics of the main and auxiliary heat exchangers;
3.   parameters of heat source (water, air and so on), its heat capacity and temperature, possibility to change its heat flow;
4.   characteristics of heat consumption; stability in heat demand;
5.   parameters of refrigerating medium, working conditions, heat capacity;
6.   heat pump operating mode, its accordance to current parameters of heat source and consumer demand, possibility to change the mode respective to outside parameters alternations.

Let us analyze the influence of heat consumption on energy characteristics of the heat pump. One of the main parameters that effect efficiency is temperature of refrigerating liquid condensation. The process of heat exchange between the refrigerant and the consumer heating system consists of the processes shown in Fig. 1:

– heat transfer from the refrigerating medium (Freon) to the inner side of the condenser pipe (q_2);
– heat conduction through the condenser wall (q_3);
– heat transfer from the outer side of the condenser pipe to the refrigerating medium in the heating system (water, air) (q_4).

These processes in condenser are described with equations:

$$q_2^c = \alpha_{in}^c \left[T - T_{in}^c \right] \cdot S_{in} , \qquad (1)$$

$$q_3^c = \frac{\lambda}{\delta} \left[T_{in}^c - T_{ex}^c \right] \cdot S_t , \qquad (2)$$

$$q_4^c = \alpha_{ex}^c \left[T_{ex}^c - T_s \right] \cdot S_{ex} , \qquad (3)$$

where T – temperature of the refrigerant, K; T_{in}^c, T_{ex}^c – temperature of inner and outer side of the condenser pipe wall, K; T_s – temperature of the working medium in the consumer heating system, K; S_{in}, S_t, S_{ex} – the area where heat transfer and conduction take place, m²; λ, δ – heat conduction coefficient, Wt/(m·K); δ – thickness of the condenser pipe wall, m; α_{in}^c, α_{ex}^c – coefficients of heat transfer for inner and outer side of the condenser pipe wall.

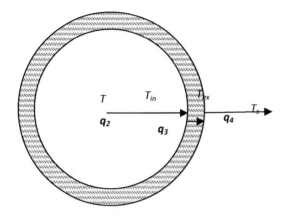

Figure 1. Scheme of heat transfer from refrigerant to the working medium in the consumer system

For example, coefficients of heat transfer in these equations could be defined as:

$$\alpha_{in}^c = 0,2 C \sqrt[4]{\frac{r \rho^2 \lambda^3 g}{\mu \theta_a l}} \cdot (Re'')^{0,55} \cdot (Pr)^{-0,33} \cdot 10^{-3} ,$$

$$\alpha_{ex}^c = 9750 \left(\frac{Gr}{2lz} \right)^{1/3} .$$

According to the law of energy conservation for the stationary stage:

$$q_2^c = q_3^c = q_4^c ,$$

$$\alpha_{in}^c \left[T - T_{in}^c \right] = \frac{\lambda}{\delta} \left[T_{in}^c - T_{ex}^c \right] = \alpha_{ex}^c \left[T_{ex}^c - T_s \right] . \qquad (4)$$

To reach the minimum of energy consumption, it is necessary to maintain the optimum temperature of condensation:

$$\alpha_{in}^c \left[T - T_{in}^c \right] = const . \qquad (5)$$

From (4) we can see that change in temperature T_s or in speed of the working medium flow in the consumer system (that effects coefficient α_{ex}^c) causes change in T_{in}^c. Thus it is necessary to control temperature and flow rate of refrigerating medium to stabilize the process of condensation (5). To solve this problem, storage tanks are used in the heating systems. But this approach requires additional capital costs and is not allow to eliminate the influence of heat consumption character on the heat pump mode and efficiency.

The same processes occur in the evaporator and cause reduction of energy efficiency of the heat pump. The temperature of evaporation has even more significant impact on energy efficiency than temperature of condensation in the condenser. There is no possibility to change temperature of the heat source or coefficient of heat transfer in the evaporator. It is a reason to take into account current temperature value of the heat source and maintain parameters of refrigerating medium respective to it. It will allow to reduce specific energy costs.

We conducted experimental and theoretical research into influence of heat consumption and supply mode on energy efficiency of heat pump was made. Research was done jointly by Esslingen University of Applied Science (Germany) and State educational Institution "National Mining University" (Ukraine) under the aegis of the National Agency for Science, Innovation and Information of Ukraine and BMBF (Germany).

The test unit for experimental study of heat pump parameters is shown in Fig. 2. Heat pump (Airconditioner) Dometic HB 2500 of "air – air" type was used.

(a)

(b)

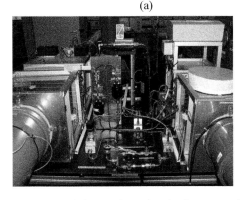
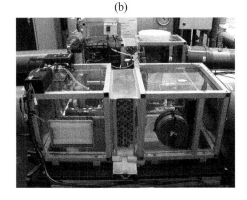

Figure 2. Test unit for experimental study of parameters heat pump respective to operating mode and disturbances: *a* – frontal view; *b* – side view.

The main parameters of working liquid (Freon R-410), such as pressure and temperature, were measured in the test unit:

- after the evaporator;

- after the compressor;

- after the condenser;

- after the throttling valve.

We also monitored the speed and temperature of incoming air (source of heat) and outcoming air (consumed heat), power consumption by the heat pump. These data allowed to calculate other characteristics of heat pump operation, such as coefficient of performance (COP) for different modes of operation and disturbances.

3 ANALYSIS OF EXISTING CONTROL ALGORITHMS OF HEAT PUMPS

Existing modes of heat pump control were studied with the help of test unit. Usually heat pumps are designed to vary heat supply by changing the speed and quantity of outgoing air. Parameters of the refrigerant in relation to operating mode of heat pump are presented in Fig. 3.

Fig. 3 presents four zones related to different modes of heat pump operation.

Zone 1 – transit process of heat pump switching on. Airflow speed reaches the required value during first seconds of operation, but it takes approximately 15 min to increase temperature of the working fluid.

Zone 2 – heating mode "3" (according to heat pump specification). Airflow speed and energy consumption are maximal.

Zone 3 – heating mode "2". Airflow speed is 2 – 4 % lower in comparison with that in Zone 2, its temperature being also 2 – 3 % lower. Power consumption of heat pump is only 1% lower than in Zone 2.

Zone 4 – heating mode "1". Outgoing air speed and temperature 6 % lower, energy consumption 2% lower compared with that in Zone 2.

We have witnessed that airflow velocity varies within quite a narrow range so that the operating mode of the pump approaches the most efficient one. However, if we take into account the drop in the incoming air temperature and the rate of its supply, we can conclude that the specific energy consumption per generated heat unit grows as the pump operation mode is other than the optimal. In this case, the minimum specific energy consumption is related to the heat pump operation at the maximum velocity of the heated air supply into the room. This fact explains the necessity to design a heat pump control mode which will adapt to changes in heat consumption with the view to maintaining the minimal possible level of specific energy consumption. The conducted experimental research confirms that as yet, there is no such control of the heat pump operation.

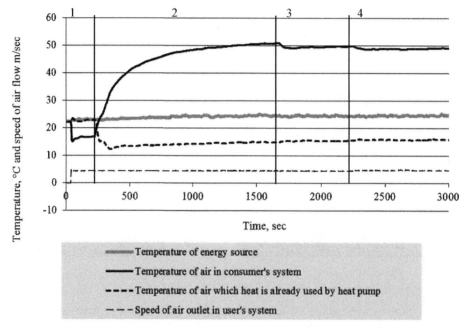

Figure 3. Parameters of heat pump refrigerating medium as a function of on operating mode. Source: own research.

Experiments were also conducted to study additional possibilities of changing operating mode of heat pump and influence of disturbances. The heat pump is designed to operate in conditions of heat consumption and supply change within the 10% range. If more significant changes in heat load occur, heat pump operates in inexpedient mode with frequent switching. It causes additional energy losses, due to momentum of heating and cooling process. As we can see in Fig. 3, it takes approximately 20 minutes for the temperature of outlet air flow in the consumer's system to reach the stable level. Therefore possibilities to change the amount and speed of inlet and outlet air flow were studied.

The dependence of heat pump COP on the speed of outlet airflow is shown in Fig. 4. Energy effectiveness changes significantly. With the increase of condenser cooling intensity, COP raises and specific energy consumption decreases. Research showed that pressure set up by throttling valve, influences thermal and energy characteristics of heat pump. The experimental

data obtained show that at high speeds of the outlet air flow, the decreasing pressure in throttling valve leads to minor raise of COP. But at lower speeds of the outlet flow, the same decreasing pressure causes significant energy losses and COP decrease.

The level of the refrigerant compression by compressor and its energy characteristics, temperature and speed of inlet air flow (or another heat source) also have significant impact on energy efficiency of the heat pump and should be taken into account in the control algorithm.

When waste heat is a source of energy for the heat pump its temperature and massflow also influence heat pump efficiency. For waste heat these parameters could be unstable and should be controlled as well.

Thus heat pump control that will provide the highest possible energy efficiency depends on numerous parameters that change in time. It is proposed to investigate modes of heat pump operation on mathematical model including all these characteristics.

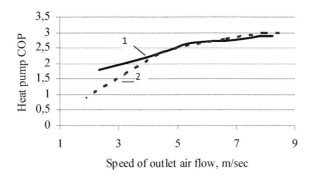

Figure 4. Experimental plot of heat pump COP as a function of heat consumption intensity: 1 – rated position of throttling valve; 2 – pressure in the system decreased by throttling valve. Source: own research.

4 MATHEMATICAL MODEL OF HEAT PUMP OPERATING MODE

The main approach used to study heat pump processes describes heat cycle in T-s diagrams and represents components of heat pump (evaporator, condenser) as point-objects changing parameters of refrigerating medium (Kato Y., Watanabe Y. & Yoshizawa Y. 1996; Sun Qiong & Zhuang Zhao-yi 2011). It does not allow to take into account possible incomplete phase transformations of the refrigerant, which occur in real processes. For example, when heat consumption decreases, incomplete refrigerant condensation or its incomplete evaporation during heat extraction from the environment takes place. These should be taken into account in calculating heat pump energy efficiency. Therefore, in operating mode study with changing heat load or waste heat amount, it is necessary to consider refrigerating medium as a two-phase medium.

To calculate parameters of the refrigerant, taking into account the actual degree of dryness, it is necessary to apply considerably more accurate description of processes in the evaporator and condenser. We propose to describe the processes by differential equations with the change of energy and mass parameters along the length.

The main assumptions adopted in the simulation:

1. The refrigerating medium is considered as two-phase medium.

2. The evaporator and condenser are considered as objects with parameters distributed along the length.

3. The throttling valve and the compressor are considered as point objects.

4. The refrigerating medium outflow in all elements of the heat pump except the condenser and the evaporator is considered adiabatic.

Further in phases of medium equations indexes j mean: 1 – steam, 2 – fluid.

To consider processes that occur in evaporator, let us analyze the equation of phase state:

$$P = \rho_1^0 R T_1 , \tag{6}$$

where P – pressure, Pa; T_1 – temperature of steam in refrigerating medium, K; R – universal gas constant.

For Freon steam and fluid a functional relation $P = f(T_1, T_2)$ can also be used.

$$\rho_1 = \rho_1^0 \alpha , \tag{7}$$

$$\rho_2 = \rho_2^0 (1 - \alpha) , \tag{8}$$

where α – volume content of steam in two-phase medium; ρ_j^0 – real density of the component, kg/m³; ρ_j – average density of the component, kg/m³.

Equations of mass conservation low for each phase of the refrigerant depend on the coordinate along the evaporator pipe (x):

$$\frac{1}{F} \frac{d}{dx} (\rho_1, w_1, F) = g_{21} , \tag{9}$$

$$\frac{1}{F} \frac{d}{dx} (\rho_2, w_2, F) = -g_{21} , \tag{10}$$

where F – area of the evaporator's pipe section, m²; w_1, w_2 – average speed of the phase flow in considered section, m/sec; g_{12}, g_{21} –

massflow density of substance under study, kg/(sec·m^2).

$$g_{21} = \beta^e \cdot \left[\frac{P_{eq}(T_2)}{\sqrt{(2\pi R\, T_2)}} - \frac{P}{\sqrt{(2\pi RT_1)}} \right] \cdot S_{21}, \quad (11)$$

where S_{21} – area of interaction between the fluid phase and the steam phase of the refrigerating medium, m^2.

Parameter S_{21} is a function of steam volume content of in the two-phase medium and depends on the regime of the refrigerating medium flow.

Equations of motion for each phase of the refrigerating medium depend on the coordinate along the evaporator pipe (x):

$$\rho_1 \cdot w_1 \frac{dw_1}{dx} = -\alpha \frac{dP}{dx} + g_{21}(w_2 - w_1), \quad (12)$$

$$\rho_2 \cdot w_2 \frac{dw_2}{dx} = -(1-\alpha)\frac{dP}{dx}. \quad (13)$$

Equations of energy conservation lows should be included in the model as well:

$$\rho_1 \cdot w_1 \frac{d}{dx}(i_1 + \frac{w_1^2}{2}) = g_{21}(i_2 - i_1 + \frac{w_2^2 - w_1^2}{2}), \quad (14)$$

$$\rho_2 \cdot w_2 \frac{d}{dx}(i_2 + \frac{w_2^2}{2}) = q_2^e, \quad (15)$$

where i_j – enthalpy of j component, J/kg, q_2^e – power of phases interaction, W.

$$i_j = c_{pj}T_j, \quad (16)$$

where c_{pj} – heat capacity of j component, J/(kg·K); T_j – temperature, K.

Processes of heat transfer in the evaporator pipe are described similar to (1) – (5):

$$q_2^e = \alpha_{in}^e \left[T_{in}^e(x) - T_2(x) \right] \cdot S_{in}, \quad (17)$$

$$q_3^e = \frac{\lambda}{\delta} \left[T_{ex}^e(x) - T_{in}^e(x) \right] \cdot S_t, \quad (18)$$

$$q_4^e = \alpha_{ex}^e \left[T_s^e(x) - T_{ex}^e(x) \right] \cdot S_{ex}, \quad (19)$$

$$q_2^e(x) = q_3^e(x) = q_4^e(x). \quad (20)$$

After evaporation process (or condensation for condenser's part of the heat pump) heat transfer coefficients significantly decrease, therefore for calculating parameters after evaporation it is necessary to condenser other values of heat transfer coefficients.

The processes in the condenser are described in the same way:

$$\frac{1}{F}\frac{d}{dx}(\rho_1, w_1, F) = g_{12}, \quad (21)$$

$$\frac{1}{F}\frac{d}{dx}(\rho_2, w_2, F) = -g_{12}, \quad (22)$$

$$g_{12} = \beta^e \cdot \left[\frac{P_{eq}(T_2)}{\sqrt{(2\pi R_1 T_2)}} - \frac{P}{\sqrt{(2\pi R_1 T_1)}} \right] \cdot S_{12}. \quad (23)$$

$$\rho_1 \cdot w_1 \frac{dw_1}{dx} = -\alpha \frac{dP}{dx} + g_{21}(w_2 - w_1); \quad (24)$$

$$\rho_2 \cdot w_2 \frac{dw_2}{dx} = -(1-\alpha)\frac{dP}{dx}. \quad (25)$$

$$\rho_1 \cdot w_1 \frac{d}{dx}(i_1 + \frac{w_1^2}{2}) = g_{21}(i_2 - i_1 + \frac{w_2^2 - w_1^2}{2}); \quad (26)$$

$$\rho_2 \cdot w_2 \frac{d}{dx}(i_2 + \frac{w_2^2}{2}) = q_2^c, \quad (27)$$

$$q_2^c = \alpha_{in}^c \left[T_2(x) - T_{in}^c(x) \right] \cdot S_{in}, \quad (28)$$

$$q_3^c = \frac{\lambda}{\delta} \left[T_{in}^c(x) - T_{ex}^c(x) \right] \cdot S_t, \quad (29)$$

$$q_4^c = \alpha_{ex}^c \left[T_{ex}^c(x) - T_s^c(x) \right] \cdot S_{ex}, \quad (30)$$

$$q_2^c(x) = q_3^c(x) = q_4^c(x). \quad (31)$$

The parts of the heat pump without actuating elements which contain only pipes with adiabatically moving refrigerating medium are located:

– between the evaporator and the compressor;
– between the compressor and the condenser;
– between the condenser and the throttling valve;
– between the throttling valve and the evaporator.

For each of these parts the conservation laws will be

$$\rho_1 w_1 F = const; \quad (32)$$

$$\rho_2 w_2 F = const, \quad (33)$$

$$P = \rho_1^0 RT_1; \quad (34)$$

$$P = \rho_2^0 R T_2, \tag{35}$$

$$P = \frac{-\alpha \rho_1 w_1^2}{2D}, \tag{37}$$

$$P = \frac{-(1-\alpha)\rho_2 w_2^2}{2D}. \tag{38}$$

Parameters of the compressor's work, power capacity of heat pump W as a function of operational mode should be also taken into account:

$$W = f(P_1, P_2, \alpha, T). \tag{39}$$

For numerical calculation, it is proposed to use finite differences method and to convert the initial system of equations for the evaporator to

$$\begin{cases} \dfrac{dP}{dx} = \dfrac{d}{dx}(\rho_1^0 R T_1), \\[2mm] \dfrac{1}{F}\dfrac{d}{dx}(\rho_1^0 \alpha, w_1, F) = g_{21}, \\[2mm] \dfrac{1}{F}\dfrac{d}{dx}(\rho_2^0 (1-\alpha), w_2, F) = -g_{21}, \\[2mm] \rho_1 \cdot w_1 \dfrac{dw_1}{dx} = -\alpha \dfrac{dP}{dx} + g_{21}(w_2 - w_1), \\[2mm] \rho_2 \cdot w_2 \dfrac{dw_2}{dx} = -(1-\alpha)\dfrac{dP}{dx}, \\[2mm] \rho_1 w_1 \dfrac{d}{dx}(c_{p_1}(T_1 - 273) + \dfrac{w_1^2}{2}) = \\[2mm] g_{21}(c_{p_2}(T_2 - 273) - c_{p_1}(T_1 - 273) + \dfrac{w_2^2 - w_1^2}{2}) \\[2mm] \rho_2 w_2 \dfrac{d}{dx}(c_{p_2}(T_2 - 273) + \dfrac{w_2^2}{2}) = q_2^e. \end{cases} \tag{40}$$

A similar system of equation should be composed for the condenser. By solving these systems, we can obtain all parameters of the heat pump for the current values of heat supply and consumption, taking into account disturbances and modes of operation of each component. Such mathematical model could be used to analyze modes of heat pump operation and to choose the mode with the highest energy efficiency relevant to current parameters.

5 RESULTS

The analysis of factors influencing energy efficiency of heat pumps and experimental data allow to conclude that energy costs could be decreased, if the heat pump is controlled taking into account operating conditions of the heat source (including unstable waste heat sources)

and the heat load. It is proposed to investigate the modes of heat pump operation using mathematical model. This model treats the refrigerant as a two-phase medium and analyses changes in physical parameters along the evaporator and the condenser. It permits to explore modes with incomplete phase transformation of the refrigerating medium, which occur in real processes.

It is planned to verify the adequacy of the proposed mathematical model to experimental data, to calculate a set of possible heat pump operating modes for different external conditions and heat load values. The analysis of modes will allow to choose the optimum control actions and work out the control algorithm.

REFERENCES

Kato, Y., Watanabe, Y. & Yoshizawa, Y. 1996. *Application of inorganic oxide/carbon dioxide reaction system to a chemical heat pump.* Energy Conversion Engineering Conference,. IECEC 96. Proceedings of the 31st Intersociety, Vol. 2: 763 – 768.

Sun Qiong. 2011. *Operating performance analysis on heating conditions of large sewage source heat pump system.* International Conference on Electric Technology and Civil Engineering (ICETCE): 1375 – 1378.

Energy Efficiency Improvement of Geotechnical Systems – Pivnyak, Beshta & Alekseyev (eds)
© 2013 Taylor & Francis Group, London, ISBN 978-1-138-00126-8

Features of filter-compensating devices application in power supply systems at coal mines in Ukraine

Yu. Razumnyi, A. Rukhlov & Yu. Mishanskiy
State Higher Educational Institution "National Mining University", Dnipropetrovs'k, Ukraine

ABSTRACT: The article deals with the results of filter-compensating device (FCD) application in power supply system at the GEROEV KOSMOSA mine (PJSC "DTEK Pavlogradugol"). The authors assert that application of controlled electric drives based on power transformers at the mine results in systematic power quality derating. The 5th harmonic FCD application is a complex solution to the problems of power quality and reactive-power compensation.

1 INTRODUCTION

In recent years the frequency controlled electric drive for high-power stationary electrical plants (namely, the fan of main ventilation (FMV), all types of hoisting units, conveyer transport) has tended to be applied in the mining industry of Ukraine, and particularly, in Mine Group Production Structural Divisions of PJSC «DTEK Pavlogradugol».

Intensive implementation of power converting equipment (i.e. power electronics) into power supply systems of mines causes a negative effect in the form of electromagnetic interferences (EMI). As a rule, they are of harmonic character, and up to the end of the last century the higher harmonics in power supply systems were regarded as the main type of electromagnetic interferences. Nowadays, due to the aging and substantial deterioration of power industry funds (70-80 % in Ukraine and Russia), problems of voltage quality are becoming more urgent. The attention is focused on voltage quality increase as far as electrical power systems of developed countries are concerned as well. It is estimated (Zhezhelenko, Sidlovskiy, Pivnyak & Sayenko 2009), the problems of power quality cost the industry, and in general, the European Union about €10 mlrd per year, while prevention measures make up less than 5 % of the sum.

2 ELECTROMAGNETIC COMPATIBILITY PROBLEM AND SOLUTIONS

Electric power as a commodity has a range of specific properties. Normalization of power quality indices (PQI) is one of the main aspects of electromagnetic compatibility problem (Zhezhelenko 2004). The basic normative document in which the requirements to quality of electrical energy in general-purpose power grids are laid down is the State Standard 13109-97 (GOST 1997).

According to this Standard, some PQI characterize electromagnetic compatibility of electrical equipment of power supply enterprise and its users at a steady-state regime regarding the peculiarities of technological process of production, transmission, distributing and consumption of electric power. They cover deflections of voltage and frequency, distortion of voltage waveform harmonicity, voltage asymmetry and voltage oscillation. To normalize them, legitimate values of PQI were set (GOST 1997).

The other PQI cover transient noise arising as a result of switching processes, storm and atmospheric phenomena and postemergency state: voltage depressions and voltage impulses, short power supply breaks. The State Standard does not specify legitimate values for them.

It should be noted that the problem of power quality in Ukrainian grids is extremely specific. For instance, in industrially developed countries of Western Europe, the linking-up of high-power nonlinear loadings which distorts grid current and voltage waveforms is permissible only in case of compliance with power quality requirements and in the presence of proper adjuster devices. The total power of nonlinear loading must not exceed 3-5 % of company's loading power. As far as Ukrainian grids are concerned, the situation is different: the linking-up of high-power unbalanced and nonlinear collectors is sporadic. The developed system of power tariff extra charges for voltage quality derating does not actually work.

The objective of this article is to analyze the situation on voltage quality in the electric grid feeding PJSC «DTEK Pavlogradugol». Special attention must be paid to the estimation of power consumption conditions at the GEROEV KOSMOSA mine, in the power supply system of which the filter-compensating device produced by electrical company "ČKD ELEKTROTECHNIKA" was installed in 2010.

A decline in higher harmonic (HG) levels within electric systems is an integral part of the general task to diminish the impact of nonlinear loadings on supply mains and to improve power quality in power supply systems of enterprises. The end-to-end solution of this task, based on application of multifunction devices, appears to be more economically feasible, than, for example, measures on current transformer upgrading. The example of such multifunction devices are resonance filters often referred to as filter-compensating devices (FCD), which apart from decline in HG levels generate a reactive-power in supply mains (Zhezhelenko, Shidlovskiy, Pivnyak & Sayenko 2009).

Resonance FCD is realized at parallel connection of LC-circuits adjusted on frequencies of single harmonic. The deficit of reactive power on substation buses in this case can be fully compensated by the batteries of FCD condensers, and the installed power of condensers is utilized up to 80–90 %. Thus, FCD are simple and low-consumption devices, that is why they are commonly used. We will consider such a device in details.

3 THE RESULTS OF CONTROLLABLE FILTER-COMPENSATING DEVICES APPLICATION IN POWER SUPPLY SYSTEMS OF COAL MINES

Simplified circuits of FCD are given in Figure 1. Figure 1, (a) shows the circuit if live condenser insulation relative to ground does not exceed phase voltage in the network: thus, performance reliability of battery improves. Regarding the ease of handling and reliability of electrical equipment layout, the circuit in Figure 1, (b) is most commonly used.

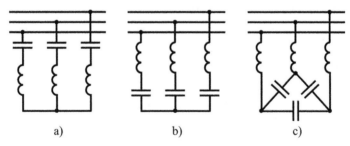

a)   b)   c)

Figure 1. The standard FCD circuits

The GEROEV KOSMOSA mine PJSC "DTEK Pavlogradugol" can be fed according to the three voltage levels: 6, 35 and 150 kilovolt (operating voltage 150 kilovolt). The coal and rock hoists are equipped with dc motors with valve voltage inverters and TEDP control system (thyristor electric drive package). Each of two 6 kilovolt bus sections is connected with the batteries of static capacitors (BSC). The ventilator of main ventilation (its power is 3,8 megawatt) with the asynchronous thyristor cascade system (ATC), and besides, FCD produced by electrical company "ČKD ELEKTROTECHNIKA" are installed at the mine. The FCD is tuned in filtration of the 5th harmonic, the capacitors power is 2×2500 kiloVAr. Simplified circuits of FCD are shown in Figure 2.

Taking into account that the GEROEV KOSMOSA mine is the first mining enterprise in Ukraine with controllable FCD applied, the experimental analysis of power quality indices was carried out under different operation conditions. The purpose of the researches was the following:

- to determine distortion levels of voltage curve harmony in case FMV works with ATC system;

- to determine decrease levels of voltage nonsinusoidality ratios and n-th harmonic components on application of the 5th harmonic FCD;

- to calculate the effect of BSC power on K_U and $K_{U(n)}$ increase caused by the grid resonance.

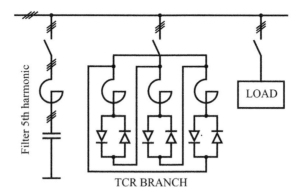

Filter 5th harmonic

LOAD

TCR BRANCH

Figure 2. The 5th harmonic FCD circuit produced by company "ČKD ELEKTROTECHNIKA"

By integrated practical study carried out at the GEROEV KOSMOSA mine on application of "FLUKE 435" voltage quality analyzer, time dependences of quality indices on nonsinusoidality (Figure 3) and on power-factor (Figure 4) have been obtained. Dependences show the efficiency of 5th harmonic FCD application.

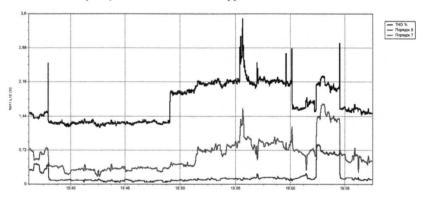

Figure 3. Dependence of harmonicity distortion factor (THD) and higher voltage and current harmonics upon the mine operation conditions

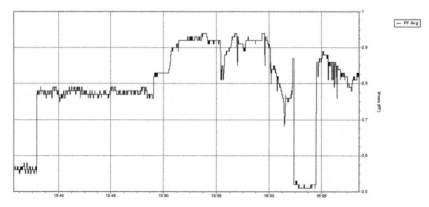

Figure 4. Variation of power-factor (PF)

Analyzing the dependences obtained, one can see that if the FCD is switched on at 15.38 PM, the value of harmonicity distortion factor drops from 1,6 to 1,5 %. The level of the 7th harmonic

67

drops likewise the distortion factor. The 5th harmonic is almost completely absorbed from the electrical power system. The insignificant indices splashes observed at 15.49 and 16.00 can be explained by the rise of resonance phenomena in the oscillatory circuit «feed-in network - static capacitors battery» (the BSC is switched on at that time interval). During the time interval 16.02-16.04 the FCD was powered-off, and the immediate increase of indices level testifies it.

The unique technical feasibility to compensate reactive power in the mine power supply, which the installed FCD produced by "ČKD ELEKTROTECHNIKA" demonstrates, should be pointed out. According to the measurements, during the time interval 15.38-16.02 the average value of power-factor (PF) used to be within the limits of 0,85, that is an excellent index for a coal mine. At the same time, in case of FCD switching off from 16.02 to 16.04, the average *PF* value dropped to 0,52 at invariable value of loading power in the mine. Thus, practical researches on power quality indices carried out by the group of scientists from National Mining University (Dnipropetrovs'k) have testified the application efficiency of controllable FCD produced by "ČKD ELEKTROTECHNIKA" if one needs to increase electric power quality indices and those of reactive power compensation in the electrical power system of Ukrainian coal mines.

To compare the results of experimental researches, calculate voltage unharmonicity indices in the electrical power system of the 6 kilovolt mine in case the FMV with SCR's control is implemented there. To show the utility operating efficiency, calculations are made as for FCD being switched off.

Basic data for calculations:
- short circuit power in case of 150 kilovolt buses: $S_{K1} = 488$ MVA;
- short circuit power in case of 6 kilovolt buses: $S_{K2} = 153,5$ MVA;
- frequency transformer power for FMV: $P_{MV} = 3,8$ megawatt, $S_{MV} = 4,5$ MVA;

- power of the motive loading of 6 kilovolt section: $S_M = 10$ MVA, $\cos\varphi = 0,8$;
- substation transformer of TDTN-25000/150 type ($S_{T1} = 25$ MVA).

The equivalent circuit relative to the initial power supply circuit regarding the GEROEV KOSMOSA mine is given in Figure 5.

– Parameters of equivalent circuit (Ivanov & Sokolov 1987):

$$x^*_{line} = \frac{S_{MV}}{S_{K1}} = \frac{4,5}{488} = 0,0092;$$

$$x^*_{T1H} = \frac{0,5\left(\Delta u_{\kappa H-L} + \Delta u_{\kappa H-M} - \Delta u_{\kappa M-L}\right)}{100} \cdot \frac{S_{MV}}{S_{T1}} =$$
$$= \frac{0,5(18+10,5-6)}{100} \cdot \frac{4,5}{25} = 0,02;$$

$$x^*_{T1L} = \frac{0,5\left(\Delta u_{\kappa H-L} + \Delta u_{\kappa M-L} - \Delta u_{\kappa H-M}\right)}{100} \cdot \frac{S_{MV}}{S_{T1}} =$$
$$= \frac{0,5(18+6-10,5)}{100} \cdot \frac{4,5}{25} = 0,0122;$$

$$x^*_{T1} = x^*_{T1H} + x^*_{T1L} = 0,02 + 0,0122 = 0,0322;$$

$$x^*_M = x''_d \frac{S_{MV}}{S_M} = 0,12 \cdot \frac{4,5}{10} = 0,054;$$

$$x^*_c = \frac{\left(x^*_{line} + x^*_{T1}\right) \cdot x^*_M}{x^*_{line} + x^*_{T1} + x^*_M} = \frac{(0,0092+0,0322)\cdot 0,054}{0,0092+0,0322+0,054} =$$
$$= 0,023;$$

where $x^*_{MV} = 0$,

$$\sin\varphi = \sqrt{1-\cos^2\varphi} = \sqrt{1-0,8^2} = 0,6.$$

– The harmonicity distortion index in case of 6 kilovolt buses:

$$K_U = x^*_c \sqrt{\frac{3}{\pi} \cdot \frac{\sin\varphi}{x^*_c + x^*_{MV}} - \frac{9}{\pi^2}} = 0,023\sqrt{\frac{3}{\pi} \cdot \frac{0,6}{0,023+0} - \frac{9}{\pi^2}} =$$
$$= 0,1 = 10\%,$$

that is greater than legitimate value in accordance with the State Standard (GOST 1997).

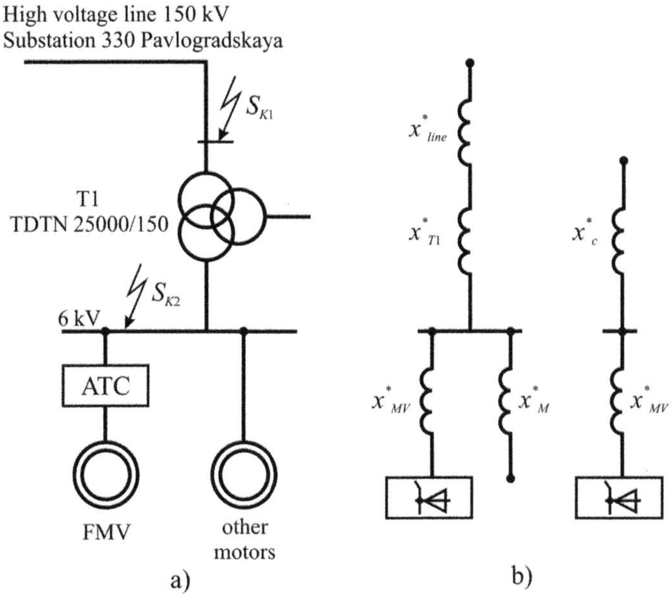

High voltage line 150 kV
Substation 330 Pavlogradskaya

S_{K1}

T1
TDTN 25000/150

S_{K2}

6 kV

ATC

FMV   other
motors

a)

x^*_{line}

x^*_{T1} x^*_c

x^*_{MV} x^*_M x^*_{MV}

b)

Figure 5. Power supply circuit (a) and equivalent circuit (b) for the GEROEV KOSMOSA mine

4 CONCLUSIONS

Thus, analyzing the results of theoretical and practical researches of power quality indices, one can assert that installation of frequency controllable FMV at the GEROEV KOSMOSA mine leads to systematic indices derating. The installation of the 5th harmonic FCD produced by "ČKD ELEKTROTECHNIKA" is the general solution to power quality and reactive-power compensation problems, that's why integrated expediency assessment of installation of such units in other mine group production structural divisions of PJSC «DTEK Pavlogradugol» is required.

REFERENCES

Zhezhelenko, I.V., Shidlovskiy, A.K., Pivnyak, G.G. & Sayenko, Y.L. 2009. *Electromagnetic compatibility of electrical supply* (in Russian). Dnipropetrovs'k: National Mining University: 319.

Zhezhelenko, I.V. 2004. *Harmonics in electrical supply systems of industrial plants* (in Russian). Moscow: Energoatomizdat: 358.

GOST 13109-97. *Electrical energy. Electromagnetic compatibility of technical devices. Standards of electrical energy quality in supply systems of general-purpose.* Issued in Ukraine, January, 1, 2004 (Interstate standard of CIS - Commonwealth of independent states), 30.

Ivanov, V.S. & Sokolov, V.I. 1987. *Modes of consumption and quality of electrical energy system of industrial plants supply* (in Russian). Moscow: Energoatomizdat: 336.

The features of energy efficiency measurement and control of production processes

S. Vypanasenko & N. Dreshpak

State Higher Educational Institution "National Mining University", Dnipropetrovs'k, Ukraine

ABSTRACT: This paper reports that by using regression analysis, the energy efficiency control at an industrial enterprise ensures accounting of structural and regime changes. The factors are determined which influence the frequency control and the duration of an interval with invariable values of planned indicators.

1 INTRODUCTION

Ensuring effective energy efficiency control of production processes at enterprises of Ukraine contributes to solving problems of its energy security, improves the ecological situation. The existing principles of energy management at industrial enterprises are based on the structural features of the production management apparatus, existing subordination of its individual links, and functional division of responsibilities between heads of departments. Introducing the responsibility for energy consumption of an enterprise in the scope of official duties of a chief power engineer, without maintaining solid relations of his activities with the process management directly at workplaces, results in the reduced energy efficiency of production. Wide involvement of heads of structural divisions as direct executers of technological operations in energy management, expansion of their authorities and strengthening of their responsibilities in the field of energy saving opens up new opportunities for the process organization on principally new approaches. Energy consumption planning directly in structural divisions of an enterprise, its comparison with actual consumption, creates all the possibilities for the continuous control of the achieved energy efficiency performance, its significant improvement in the management process. Energy management systems of industrial enterprises, where new management principles are used, are well-known in the international practice and they have high efficiency. This study outlines the basic principles of such systems, determines the content and sequence of interaction of their

individual components, and gives examples of their implementation in mining.

2 GENERAL PRINCIPLES OF ENERGY MANAGEMENT SYSTEMS

The way of energy consumption control, inherent in the energy management systems, is sufficiently effective. It can be used to control energy use of any enterprise. General principles of these systems are well-known. The basic idea consists in the personal responsibility of heads of individual divisions of the enterprise for energy consumption of these divisions. Specific implementation of the system is possible when taking into account the features of the enterprise for which such a development is being performed.

A characteristic feature of the control and normalization systems is the continuity of the control and management processes. This approach to management has generalized the experience of developed countries in achieving a maximum effect when carrying out the energy saving policy of the company. This has become possible thanks to the clear structure of the energy management system, determining concrete actions of its separate links and solving the problems of information and scientific support.

The block diagram given in Fig. 1 explains the mechanism of action of the system.

Efficient energy management of an enterprise is possible when the energy consumption minimization will be achieved directly at workplaces, that is, where this energy is used. To realize such kind of management it is necessary to ensure the responsibility for energy efficiency directly in structural divisions of the enterprise.

Thus, the enterprise structural divisions solve energy saving problems independently, and their heads are responsible for the results of such activities. In the energy management system, technological objects of selected divisions are the objects of management where it is necessary to ensure high energy efficiency of processes. Several objects are usually selected at the enterprise.

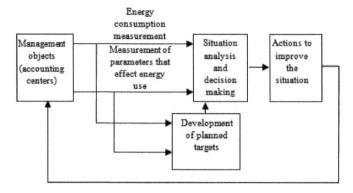

Figure. 1. Structure of the energy management system

The object of management can be a technological line, a set of the same equipment that has a compact location and production areas of an enterprise. It is important that energy consumption of these objects be significant, and there be opportunities for its measurement. In addition, it is necessary to choose people from among the management personnel who will be responsible for energy efficiency of the object. The objects selected in the structure of an enterprise are called centers for energy accounting (CEA). In the centers for energy accounting, it is obligatory to determine regime parameters that affect energy consumption and to choose the basic ones from among them (those parameters that influence energy consumption in greater measure).

Gathering information about the level of energy consumption and the parameters that determine it is carried out in the course of time. The time should be sufficient to get the required amount of statistical information and determine, on its basis, standard planned targets (of regression dependence based on the results of previous measurements).

In the current period, energy consumption control is realized by energy consumption centers on the basis of current measurements (ones a day, -week, -month or some other period of time). These indicators are compared with the planned ones. On this basis, the conclusion is made about the level of energy consumption at the object. It is important that the energy efficiency indicator of a technological process be controlled which is determined by the energy use for achieving results of this process. Therefore the parameters that affect energy consumption (the amount of output) are measured. Then the reports are made, which are analyzed by special structures (energy consumption control departments). These departments, together with the enterprise governance structures, make decisions that help to improve the situation, control the implementation of decisions adopted by energy consumption centers.

Obviously, the situation at a management object periodically changes. It is connected with the use of new technologies, modern high-performance equipment and implementation of energy saving measures. Such changes should help to reduce specific energy consumption in centers for energy accounting. It is therefore necessary to modify planned targets as well. This should be expressed in the formation of planned targets oriented on the change of the situation at the management object.

Thus, the functioning of the energy management system should be considered as an ongoing process that can take into account changes occurring at the object, specific features of its operation, variability of regime parameters over time. That is why we are talking about the normalization of energy as "an indicator that can take into account different circumstances of energy use at the object" (Prahovnyk 2001).

It is clear that in a certain period of time, the system will ensure minimal energy consumption at the object, due to specific conditions of its work. This is a very important factor, as the values of specific energy consumption norms (targets) are formed directly at the enterprise. They do not remain constant but change depending on the situation. These norms are obtained in real time and are used for continuous analysis and making decisions. In addition, there is a system that is able to improve these values.

The characteristic feature of the considered structure is the management system isolation. Indeed, there is an inverse relationship between certain actions aimed at improving the situation and a management object. Effective mechanisms should be developed to implement tight feedback. Its absence will result in the system failure. It is necessary to realize that not an automated management system is meant, but a system that provides clear interaction between people involved in its work.

To sum it up, let's select several important points concerning the functioning of energy management systems:

− measurement of different forms of energy and fuel, as well as of parameters that affect these levels;

− development of planned targets of energy consumption and energy efficiency rating in separate centers for energy accounting and within the enterprise;

− receipt of information on energy consumption of an object in an appropriate form for analysis;

− development of energy efficiency measures and mechanisms for their implementation.

3 CENTERS FOR ENERGY ACCOUNTING OF INDUSTRIAL ENTERPRISES

An important stage in the development of energy management systems is identifying the centers for energy accounting. Proceeding from the fact that a CEA is an object of management it is necessary to pay attention to the procedures of its formation taking into account the requirements it should meet. These are:

− significant energy consumption of the structural division of an enterprise, on the basis of which a CEA is formed;

− accounting of the total energy consumed by centers for energy accounting;

− appointment of a person who will be responsible for energy efficiency of a CEA.

In the centers for energy accounting, selected within the enterprise, a greater part of the energy consumed by the enterprise should be controlled. It is achieved by creating centers for energy accounting on the basis of energy-intensive consumers. Thus, a relatively small number of centers for energy accounting makes it possible to control a significant part of energy. It is desirable that the results of work of centers for energy accounting be easily controlled. It will help to analyze and improve the energy efficiency indicators of the division. The most successful example of a separate CEA selection is its formation on the basis of a production line with a controlled number of products obtained at regular intervals. Then it is possible to compare energy consumption with the volume of output, that is, to assess the process efficiency. How to identify the biggest energy-intensive consumers? It is necessary to analyze annual energy balances of the enterprises where the total energy consumption is distributed between separate divisions as well as consumers. If energy balances of the enterprise are not made, it is possible to analyze passport data of basic (energy-intensive) equipment. Its location on the territory of the enterprise will give information on the CEA possible location. As an example, let us examine energy balances of coal mines.

It should be noted that the main components of energy consumption in mining have been identified. Their list is well-known. However, according to separate items of energy balances, the ratios of energy consumption at different enterprises differ from each other. It depends on the conditions of coal mining. The energy balances of mines are presented below to show the most important items of energy consumption and percentages between them.

Indicative energy consumption limits (%) are as follows (Pivnyak 2004):

− mining and preparatory works	5 – 20
− transportation	2 – 10
− lifting	10 – 30
− pumping	5 – 40
− ventilation	20 – 30

– compressors	10 – 60
– lighting	2 – 5
– own needs	10 – 20

In mines with flat occurrence of seams, the share of the energy consumed by compressor sets is negligible. At the steep occurrence of seams the compressed air equipment is widely used. Here the relevant proportion of energy consumption increases (up to 60%).

Let's consider the energy balance (%) of underground consumers performed, for example, for the "Yuzhnodonbasskaya number 1" mine (Western Donbass):

– main drainage installations	
– underground transport:	
• electric locomotives	0,9
• trunk conveyors	10
• people and goods transportation	0,9
– mining area	10,5
– preparation area	7,2
– other consumers	0,9
Totally (the total consumption of mine)	33,7

In general expenses for the acquisition of fuel and energy resources in mines with flat fall of seams, the cost of electricity is 81-90%, of coal for own use (boiler-house) 5-15%, of water 3,8-5,3%, and in the mines with steep fall of seams, accordingly 80–87, 6–13, 4–7. This confirms a significant share of electricity in the structure of expenditures.

For the mine whose indicators are given in the international standard "Methods for determining energy consumption norms by mining enterprises" (GOST 30356-96), the structure of electric energy consumption (%) is as follows:

– mining areas	5,26
– preparation area	1,29
– underground transport	5,6
– air conditioning	10,88
– lifting	13,32
– pumping	14,28
– ventilation	17,13
– technological complex of surface	3,75
– compressed air production	2,46
– other electric consumers	20, 58
– lighting	0,69
– energy losses	4,76

From this it follows that existing and recommended indicators of power consumption differ greatly. The structure of energy balance must be changed by reducing the costs of energy-intensive components: pumping, lifting and ventilation of mines. These components of energy balance correspond to the full list of mining operations. The degree of unaccounted energy costs is low. This fact is very important for the correct identification of the place for a CEA where the main part of consumed energy must be taken into account.

It is an important fact that the given percentage numbers of energy balances are considerable. This proves that it could be expedient to control these cost items by separate CEA, that is, each cost item should be matched with its center. How feasible this approach is to the selection of centers can be proved by investigating other conditions specific to the CEA. Variation in percentages, observed in energy cost items of the given energy balances, can not influence the general conclusion about the need of controlling relevant cost items of energy consumption within a separate CEA. In addition, a CEA will exercise sufficiently accurate accounting of energy, which will make it possible to form energy balances of mines for any period of their work.

The next requirement for CEA consists in the need of accounting the total energy consumed by the center. This requirement is conditioned by full responsibility of the CEA authorities for the volumes of energy consumption. These are all forms of energy and fuel used by the center. Accurate accounting of power consumption causes no doubt of the fact that the energy is consumed directly in the CEA. This is the basis of implementing full responsibility for the level of energy consumption in the CEA. Localization of energy accounting in CEA requires a detailed analysis of energy distribution systems existing at an enterprise. The distribution systems of electricity, gas, water, steam are meant. If the structure of a distribution system ensures energy supply of several CEA, then to account energy consumption it is necessary to install meters for each of them. That is why the CEA formation process, as a rule, involves installation of additional meters and analysis of existing distribution systems makes it possible to determine concrete locations of their installation.

4 ENERGY EFFICIENCY CONTROL

It is proposed to exercise energy efficiency control in CEA on the basis of the developed regression models. The features of such models formation as well as the principles of energy efficiency control are described below.

Energy consumption in CEA can be considered as random variables that change over time. Values of energy consumption depend on a number of technological parameters which also vary randomly. Functioning of energy efficiency control systems involves comparing the actual values of energy consumption in CEA with the planned ones. The actual values can be measured directly and the planned ones are obtained from the regression dependence, based on the results of prior observations (measurements). Thus, mathematical models based on the regression dependences serve as a basis for the formation of energy targets. Besides, the energy consumption control requires a correlation analysis which will make it possible to assess the connection between energy consumption values and parameters that define them.

First of all, it is necessary to substantiate the periodicity of synchronous energy consumption measurements (by meters), and also the corresponding values of technological parameters in CEA. Such periodicity can be determined by the duration of:
 – batch production;
 – a shift;
 – a day;
 – a week;
 – a month.

Periodicity of measurements depends on the dynamics of energy consumption processes. If energy consumption changes over time significantly, measurements should be carried out more frequently.

There are different methods of gathering information necessary for the construction of the regression dependence:
 – manual;
 – with the help of computer systems for energy consumption accounting.

Each of these methods has its advantages and disadvantages. The manual measurement method involves direct participation of staff in the registration of meter readings installed in CEA. The disadvantage of this method lies in possible human errors made in the process of meter registration. However, this method is simple and most acceptable in case when measurement periodicity is significant.

Computer accounting systems are widely used for energy consumption accounting. They usually have a standard set of functional possibilities, ensure high accuracy of measurement of parameters with a given time interval. Meter readings can be transferred by informational communication channels directly to the computer that forms the regression dependence. Basic disadvantages of computer accounting systems lie in their high cost and in the necessity of their adjustment that often requires solving numerous problems of combining different software products. In addition, the automatic registration of technological parameters is not always possible.

Let's consider the task of building the regression dependence between two parameters. The selection of experimental data is formed:

$$(x_1, y_1) (x_2, y_2) \dots (x_n, y_n), \tag{1}$$

where x_i – an independent technological parameter; y_i – dependent parameter (level of energy consumption). Sometimes, in the process of collecting statistical information for the construction of the regression model, situations appear when accuracy of individual observations (measurements of parameters) is doubtful. This may occur as a result of:

• conventional random fluctuations conditioned by the nature of analyzed general summation;

• a breach of obtaining experimental data (for example, synchronicity of determining the values x_i, y_i);

• errors that occur during the registration of meter readings.

In case when observation results connected with the nature of the phenomenon differ, they cannot be excluded from consideration. To exclude the observation results, caused by the two mentioned factors, the formal methods are used. In this case, a descriptive data analysis or statistical criteria are used. Statistical criteria are based on the average values and mean square deviations of random variables y_i Excess levels of unlikely deviations require their exclusion from the set of the given measurement.

In the conditions of the well-known boundary values of the parameters x_i, y_i (which is typical, for example, for levels of production and consumption of a coal mine), a meaningful analysis can be used. Its idea is that each of the values of the independent variable x_i and dependent variable y_i is compared with the limit values of these parameters x_{min}, x_{max} and y_{min}, y_{max}. The values which do not satisfy such conditions must be excluded from the sample:

$$x_{min} \leq x_i \leq x_{max} \qquad i = 1 \ldots n ; \qquad (2)$$

$$y_{min} \leq y_i \leq y_{max} \qquad i = 1 \ldots n . \qquad (3)$$

If any of the values x_i or y_i exceeds the specified limits, either x_i or y_i values are excluded from the sample (although one of these parameters may not exceed those limits). It is important to choose correctly the limit values of x_{min}, x_{max}, y_{min}, y_{max}. However, according to the measurement results of the parameters x, y for coal mines conditions, their variation is small, significant deviations from the average values are observed very rarely and they may be associated with errors made when obtaining meter readings or introducing them in the database. In this case, the logic of selecting the sample x_{min}, x_{max}, y_{min}, y_{max} should be as follows: "these parameters, in principle, can't exist." After excluding variables from the sample it is possible

to go to the further processing of experimental results.

The results of tests that are fixed in a statistical sample must be homogeneous. Testing homogeneity is one of the most important conditions for the correct application of statistical observation methods. To ensure testing homogeneity each of the series must be carried out in stable conditions (using the same equipment, conducting experiments in the shortest possible time and remembering that many factors change over time). After obtaining the experimental data it is necessary to analyze carefully the testing conditions.

Checking the homogeneity of observations can be made by statistical methods. What you need to know is the distribution of a random variable. In general terms, the problem of homogeneity of observations is formulated as follows: if sample elements are compatible with the hypothesis that they are obtained from one and the same general summation. It is possible to use well-known comparison methods of variances and mean values.

Then the regression equation is considered as a dependence between the average values of the parameter y (a_y) and the parameter x [$a_y = \varphi (x)$]. Since the average values of a_y are valid values of observed quantity (excluding a random factor), the regression shows the dependence between these values devoid of random layers.

Under the limited number of observations it is not possible to build the dependence $a_y = \varphi (x)$. It goes about the construction of an approximate regression $\hat{y} = f(x) \approx a_y$, where $\hat{y}$ – assessment of average value of a_y.

Figure 2 shows a linear dependence between the $\hat{y}$ and x parameters which are based on the observation results.

$$S = \sum_{i=1}^{n} \left[y_i - f\left(x_i\right) \right]^2 \qquad (4)$$

The dependence $\hat{y} = f(x)$ is based on the principle of least squares which ensures minimal scattering of S values of y_i around the $f(x)$ function:

The kind of dependency is approximately determined by the nature of experimental data location. The linear regression $\hat{y} = \alpha + \beta h$ is widely used. This kind of dependence is most prevalent in existing energy management systems.

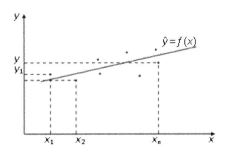

Figure. 2. Linear regression dependence $\hat{y} = f(x)$

The regression coefficient (Vypanasenko 2005)

$$\beta = \frac{n\sum x_i y_i - \sum x_i \sum y_i}{n\sum x_i^2 - \left(\sum x_i\right)^2}. \tag{5}$$

The value β is often expressed by means of the sample variances Sx, Sy and correlation coefficient r:

$$\beta = r\frac{Sy}{Sx}; \tag{6}$$

$$Sy = \sqrt{\frac{1}{n-1}\sum(y_i - y_0)^2}; \tag{7}$$

$$Sx = \sqrt{\frac{1}{n-1}\sum(x_i - x_0)^2}; \tag{8}$$

$$r = \frac{\sum(x_i - x_0)(y_i - y_0)}{(n-1)SxSy}; \tag{9}$$

$$x_0 = \frac{1}{n}\sum x_i; \tag{10}$$

$$y_0 = \frac{1}{n}\sum y_i. \tag{11}$$

The value of constant component is obtained from the equation

$$\alpha = \frac{\sum y_i - \beta\sum x_i}{n} \tag{12}$$

It is necessary to make emphasis on some features of connection between energy consumption and target parameters characteristic of coal mines. These features incline us to use a simple linear dependence of the first order which will consider the energy consumption depending on the target indicators characteristic of concrete CEA. Let us consider these arguments:

• a linear regression should be applied in experiments where the law of "uniform accumulation" is true, that is, it is known that y is connected with the change of x but does not depend on what "quantity" of the x parameter has accumulated. This very law is acceptable for the existing dependence between energy consumption and parameters, on which it depends (in the first turn, on the volume of mining production);

• boundaries of parameters changes that affect energy consumption at a particular CEA are usually insignificant. This is explained by the constant rhythm of coal enterprises where the work is clearly directed at achieving the desired result and it is coal. Therefore in steady regimes of a mining enterprise the drivings vary slightly. With minor limit changes of an argument, the linearization of even essentially nonlinear dependence $y = f(x)$ leads to small errors, that is, it is foreseen that a linear dependence is typical for a narrow range of the x argument variation;

• a linear regression dependence has clear, unequivocally interpreted indicators that characterize the degree of dependence of one random variable on another (coefficients of correlation and determination). Mathematical expressions that determine the coefficients of regression dependence are simple. The linear dependence makes it possible to calculate the expected energy savings and predict the level of energy consumption.

The results of energy efficiency control in CEA should be presented in reports. This ensures the availability and reliability of the obtained information. The results of comparing actual and planned energy consumption indicators can be shown in the form of graphs, tables and diagrams. Therefore it is necessary to develop a computer program (Dreshpak 2012) that will help to control energy efficiency during certain periods of time. The periodicity of reporting in CEA is connected with the accepted terms of registration of meter readings and technological parameters. New data in charts, tables, diagrams appear immediately after these measurements.

To realize efficient operation of the energy management system it is necessary to create within an enterprise the structure that will make appropriate decisions, based on an analysis of the situation in CEA, and control their implementation. An approximate structure of the monitoring group responsible for energy consumption control is as follows:

- head of a company;
- head of a CEA;
- energy manager;
- chief accountant;
- head of technical department.

The tasks of the group are:
- co-ordination of actions in the field of energy efficiency within an enterprise;
- control of planned energy efficiency measures;
- assistance to energy manager in the performance of his functions.

From the enumerated tasks of the monitoring group it follows that the process of energy consumption control is carried out at the level of the enterprise authorities with the attraction of all those who are responsible for energy efficiency in separate CEA as well as energy manager as an important link in the chain of this system. Since it goes about the financial responsibility of CEA for energy efficiency indicators, a monitoring group should include the chief accountant of a company. The head of technical department provides technical evaluation of proposed energy efficiency measures.

As a rule, the head of the group organizes weekly briefings as well as monthly production meetings to discuss the situation in the center. Herewith, the results of work for a week, month or quarter are analyzed, the planned energy efficiency measures are accepted for implementation, the tasks are distributed among the members of the group and the task progress is controlled.

The functioning of the energy management system provides a close link between the energy consumption monitoring group and production divisions of an enterprise (CEA). Labor collectives should be aware of the situation in divisions and actively participate in solving their problems. Periodic discussions of the situation directly in labor collectives should be aimed at the identification of causes of irrational energy and fuel use and at the development of specific measures to improve the situation. This activity is directed and controlled by the CEA authorities.

5 CONCLUSIONS

Energy efficiency control of separate CEA is realized in energy management systems by means of regression models. A regression line is a boundary of division between satisfactory and unsatisfactory performance. Herewith, a degree of difference between actual and planned energy consumption is controlled. In the conditions characteristic of a coal mine it is advisable to focus on linear regression dependence.

REFERENCES

Prahovnyk, A. 2001. *Energy consumption control and normalization* (in Russian). Materials of the conference "Energy consumption control". Kyiv: Alliance to Save Energy: 387 – 397.

Pivnyak, G., Shkrabets, P., Zaika, V. & Razumnuy, U. 2004. *Systems for efficient energy supply of coal mines* (in Russian). Dnepropetrovsk: National Mining University: 206.

Vypanasenko, S.I. 2005. *Centers for energy accounting in energy management systems of a coal mine* (in Russian). Bulletin of the National Mining University, 10: 89 – 94.

Dreshpak, N.S. 2012. *Energy efficiency measurement and control of production divisions of an enterprise* (in Russian). Mining Electrical Engineering and Automation, Issue 88: 139-143.

Energy Efficiency Improvement of Geotechnical Systems – Pivnyak, Beshta & Alekseyev (eds)
© 2013 Taylor & Francis Group, London, ISBN 978-1-138-00126-8

Rational modes of traction network operation of transport with inductive power transfer in the coal industry

Ye. Khovanskaya & A. Lysenko
State Higher Educational Institution "National Mining University", Dnipropetrovs'k, Ukraine

ABSTRACT: In the article the features of design of the modes hauling networks are considered with the inductive transmission of energy. The algorithm of calculation of parameters of hauling network is considered. Advantages this type of transport are shown. The ways of realization of rational modes are certain. The method of determination of losses of energy is offered taking into account the features of electric networks. It is shown that the values of losses in hauling networks depend on the initial parameters of starting of hauling transformer of frequency. The task of finding of rational parameters is represented as multicriterion optimization task. The methods of finding of optimum decision are analyzed. Conclusions are done about the prospects of further researches of the modes of hauling networks of the high frequency.

1 INTRODUCTION

Conveyor transport is used on a broad scale in underground coal mine development in Ukraine. Accumulator locomotives with coal wagons are also common there. But taking into account the large number of mines which are hazardous due to the gas and dust, and moreover, the number of accidents in mining industry, we should admit that progress in the field of improving and developing the energy-efficient modes of operation for transport system with inductive energy transfer seems to be promising. A characteristic property of such transport is the absence of direct connection between the current collector and a traction supply network, which makes it intrinsically safe and applicable in mines where occupational hazard is observed.

The researches in the field of inductive energy transfer, its modernization and application, aimed at solving similar tasks, have been witnessing surges of interest on this issue at the National Mining University for many years (Dnipropetrovs'k, Ukraine) (Pivnyak 1990). Thyristor power supplies, traction supply network design, electrical equipment of electrical locomotives have been developed. Practical implementation of the promising transport leaves much to be desired, so modern CAD-tools and high level programming languages are used to simulate modes of operation of the complex to find out which modes are energy-efficient.

The purpose of this paper is to study the mathematical apparatus to search for rational (optimal) in terms of power consumption modes of traction supply network. Such formulation of the objective is urgent, as the cost of energy resources is continuously increasing all over the world, and the coal industry is a priority in the strategy of economic development of Ukraine.

2 POWER LOSS IN TRACTION NETWORK

In an earlier study, it was noted that the greatest loss of power occurs precisely in the traction supply network, as the specificity of the transport operation requires constant maintenance of the current in traction supply network of 150 A. These conditions lead to the fact that regardless of whether traction supply network is loaded by electrical locomotive or not, a fair amount of the power loss is detected. However, despite this, the energy performance, including power loss, until recently, received little attention. Nevertheless, the evaluation of these parameters allows us to determine ways to reduce the power loss in the traction supply network by regulation, in particular, the launch process in the frequency converter. Therefore of a particular interest is the power loss at a triggering mode of the traction-feeding transformer (TFT). Triggering event has previously been considered in two aspects:

- In terms of the processes taking place directly in the TFT;

- In terms of impact of the start-up range of the converter on the input parameters of traction supply network.

This approach allowed us to calculate the operational parameters of the system TFT-traction supply network at a triggering mode to ensure a smooth transition of traction supply network to steady state mode (Pivnyak & Zrazhevskiy 2004). Conducted on the basis of these calculations, the adjustment algorithm for triggering TFT makes possible to avoid overvoltage in traction supply network, leading to failure of the capacitors intended for longitudinal capacitive compensation.

But such an algorithm can't be considered fully satisfactory in terms of energy consumption until it has been substantiated, because the obtained results still lack such energy datum as loss of power. Given in (Pivnyak 1990; Kalantarov 1986) curves show the area restrictions of input parameters of traction supply network at a triggering mode. The objective of this study was to determine/detect this area more precisely by analyzing the power loss that occurs at different possible combinations of triggering mode parameters. The solution of this problem is associated with certain difficulties. These difficulties are due to peculiarities of traction supply network, consisting primarily in the distribution of its parameters, as well as a large number of capacitors of longitudinal compensation.

Power loss in the line conductor is given by:

$$\Delta P = I^2 r_0 l \tag{1}$$

where I - RMS current in the line conductor;
r_0 - complete resistance of loss in the traction supply network;
l - length of the line.

At a start-up the value of current in cross sections of traction supply network varies due to the wave properties which the network can manifest, therefore the loss of power in this case must be determined taking into account operational parameters distribution in the network (Khovanskaya & Lysenko 2005):

$$\Delta P = r_0 \int_0^l (I(x))^2 \, dx \tag{2}$$

The complexity of using the last expression for traction supply network is the inability to obtain an analytical dependence of the current on the coordinate of cross section in traction supply network. To measure the current value in the cross sections of real traction supply network is also very difficult. Therefore it is permissible to use the results of numerical experiments which

were obtained by means of the mathematical model. From the theory of experiment planning it is known that the measurements can be considered as random variables, the dependences between which one may get by using regression analysis.

3 ANALYSIS OF THE METHODS AIMED AT PLOTTING REGRESSION DEPENDENSES

It is known that the **regression analysis** is a method of statistical analysis of the dependence of random variable y on variables xj (j = 1,2, ..., k).

Regression analysis examines the relationship between one variable, called the dependent variable or characteristic and a few others, which are called independent variables.

This relation is presented in a mathematical model, i.e., the equation that relates the dependent variable (y) to the independent (x) with due regard to a set of appropriate assumptions.

Since the purpose of regression analysis is to identify the influence of the value of x on y, the latter also is called a resultant factor, and the variables x are factors that affect a resultant factor.

Regression analysis is used for two reasons:
- Firstly, the description of the relationship between variables help determine the existence of possible cause-effect relationship;
- Secondly, the obtained analytical relationship between the variables provides for candidate values of the dependent variable, if values of independent variables are known.

It is usually assumed that random variable y has a normal distribution law with a conditional assembly average $\tilde{y}$, which is a function of the arguments xj (j = 1,2, ..., k) and a constant independent of the arguments of variance σ^2.

In regression analysis, the form of the regression equation is predetermined by the analysis of the physical nature of the phenomenon and observational results.

The most common are the following forms of regression equations:
- strictly linear multivariate:
$$\tilde{y} = \beta_0 + \beta_1 x_1 + \beta_2 x_2 + ... + \beta_k x_k ;$$
- polynomial:
$$\tilde{y} = \beta_0 + \beta_1 x_1 + \beta_2 x_2^2 + ... + \beta_k x_k^k ;$$
- hyperbolic:

$$\widetilde{y} = \beta_0 + \beta_1 \frac{1}{x} \ ;$$

- exponential:

$$\widetilde{y} = \beta_0' x_1^{\beta_1} x_2^{\beta_2} \dots x_k^{\beta_K} \ .$$

The exponential regression equation can be transformed into a linear equation relative to parameters β_j by means of finding the logarithm.

The choice of the relation/dependence form is hampered by the fact that, using mathematical apparatus, theoretically the relationship between the characteristics can be expressed as a number of different functions.

The choice of equation type is complicated by the fact that for any form of dependence the equations set is selected, which to some extent will describe these relationships. Initial choice of a specific regression equation is based on the analysis of preceding researches or is predetermined by the review of similar studies in related disciplines. Since the regression equation is aimed at describing in numerical terms the relationships/dependences, it should vividly reflect the actual relations established between the studied factors.

The most appropriate way to determine the form of the initial regression equation is the **method of different equations enumeration/ trial-and-error method**. The essence of this method lies in the fact that a large number of (models) regression equations selected to describe relationships of a phenomenon or process, is implemented on a computer using a specially developed algorithm of sorting followed by a statistical test, mainly on the basis of the Student t-test and Fisher F ratio. The method is rather time-consuming and its application is justified mainly in the study of social and economic phenomena, so for the purposes of this paper its use is impractical.

The **method of expert evaluations** is used as a heuristic analysis of the main macroeconomic indices, which form a single international system of calculations, it is based on an intuitive and logical assumptions, content qualitative analysis. A peer review is analyzed on the basis of calculations and careful assessment of non-parametric indices of interdependence: rank correlation coefficients of Spearman, Kendall and concordance. This method is not applicable to this area of research.

Step regression analysis is widely used. The method of step regression consists in sequential computation of factors in the regression equation and the subsequent verification of their importance/values. Factors alternatively enter into the equation by so-called "direct method". Value verification of the entered factor helps determine how much the sum of squared residuals decreases and the value of the multiple correlation coefficient increases. Simultaneously the reverse method must be used, i.e., factors that have become insignificant by Student t-test are excluded. The factor is considered to be insignificant if entering into a regression equation it changes the value of regression coefficients only but can't reduce the sum of squared residuals and increase their value. If the model involves a corresponding factor variable and the value of multiple correlation coefficient increases while the regression coefficient remains the same (or changes negligibly), then this attribute is significant and its presence in the regression equation is compulsory. This method is also quite time consuming and impractical to use.

To solve the problem of smoothing the experimental dependence curve in the case where the dependence $y = f(x)$ is known prior to the experiment, the calculation method is usually employed, it is called as the "**method of least squares.**"

4 SIMULATION OF START UP MODE PARAMETERS

The essence of the method of least squares in the solution of the given problem is the following:
- we divide the transition process time into multiple/several values and then, for each one we obtain n values of currents $i_1, i_2, \dots, i_n$, corresponding to the values $x_1, x_2, \dots, x_n$ of cross-sections in the line cable;

Regression dependence of i on x can not coincide with the experimental values i_i in all n points. This means that for some or all points the difference:

$$\Delta_i = i_i - f(x_i) \tag{3}$$

will be distinct from zero.

It is required to select the function parameters so that the sum of the squares of the differences is the lowest, i.e. to minimize the expression:

$$z = \sum_{i=1}^{n} \Delta_i^2 = \sum_{i=1}^{n} \left[i_i - f(x_i) \right]^2 \tag{4}$$

Thus, the method of least squares is considered to be the best one for approximation of analytic functions to the experimental dependence, if such a condition as minimum sum of squared deviations of the desired analytic function of the experimental dependence is observed.

• then the experimental data for different initial conditions (I_{begin}, t_{begin}) are obtained on the mathematical model ;

• the coefficients of the regression line are determined and the regression relationship is plotted in such a way that the accuracy of the approximation is the largest;

• obtained dependence I = f (x) we use to determine the power loss in traction line at a time by the formula (2);

• we plot power losses against the start time of the process at various output parameters TFT (on Fig. 3-6).

Thus, to find the functional dependence of the current on the network cross-section coordinate it is permissible to use regression analysis, in particular, the method of least squares is employed.

Initial data sets are formed on the basis of the known dependence of the output current TFT (on Fig. 1), according to which the latter gradually increases from zero to some intermediate (initial) values, and then, to steady-state value. The current rise to the initial value happens within a time interval, which can vary in the range of 200 ... 2000 microseconds. Further increase in the output current of the frequency converter to the final value is analyzed for the time period of 200 000 microseconds, provided that proper operating conditions for electrical equipment are ensured. Output current TFT is an input parameter for the traction supply network and, therefore, it affects its mode settings. The algorithm for calculating the power loss in the traction supply network at the starting process with due regard to the changes in the output parameters TFT is shown on Fig. 2.

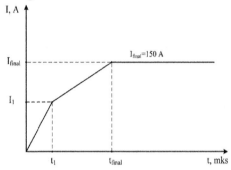

Figure 1. Changing the output current TFT during the start-up at unloaded traction supply network

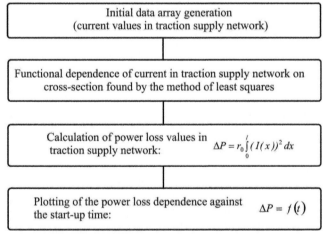

Figure 2. The algorithm structure to determine the level of loss in the traction supply network at start-up

Fig. 3 shows the influence of the initial increase of the current I_1 at a fixed time t_1 of its rise on power loss in the traction supply network. As can be seen, at any moment of a starting process the power loss increases with the initial current rise before the current reaches a steady value.

82

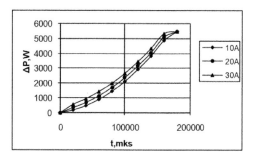

Figure 3. Dependences of the power loss values on the initial rise of the current TFT and time of the transient process

Fig. 4, 5 and 6 show the effect of the time required for current to take initial values (t_1 = 200, 1000, 2000 mks) upon the loss of power at a constant increase of initial output current TPCH I_1. It is obvious that power loss increases at all values of the time required for the initial current rise. For chosen options t_1, being equal respectively 200, 1000 and 2000 microseconds, the tendency of power loss excursion is observed as soon as current characteristics get the section of smooth loading portion. Thus for the same moment of initial phase start-up procedure and constant I_1, power loss can vary due to the time of initial increase of the current t_1, and the loss is greater, the smaller value of t_1 is. When the current takes value I_1, then in the interim of the next start-up procedure the power loss virtually remains unchanged. These graphs show that abrupt enhancement in power loss occurs alongside with a sharp increase in the output current of the frequency converter from zero to I_1 during the time interval up to 200 microseconds. The increase in the time interval t_1 at a constant value of I_1 reduces power loss in power supply traction network.

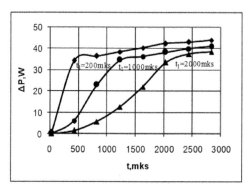

Figure 4. Power loss in the power supply traction network at startup when the time rise is altered for the output current TFT of I_1 = 10A

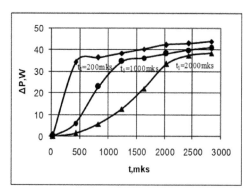

Figure 5. Power loss in the power supply traction network at startup when the time rise is altered for the output current TFT of I_1 = 20A

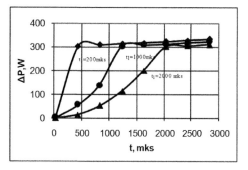

Figure 6. Power loss in the power supply traction network at startup when the time rise is altered for the output current TPCH of I_1 = 30A

The curve genus behavior of power loss is relevant to the wave properties of the line, because the currents in the sections of the power supply traction network at startup of TFT depend on the nature of change in wave impedance of traction network. While the line is not charged, the wave impedance changes more slowly than the input current, so in order to ensure minimum power loss in the power supply traction network it is necessary to determine the optimal output parameters TFT (the initial current step-up and the rise time), which would provide real effective operating conditions of the system, power supply traction TFT.

Moreover simulation model with initial parameters in terms of overvoltage observed at the beginning of the line has been considered.

For the analysis of startup processes a model for transient behavior was used (Pivnyak 1990), by which the options of start-up mode when the values of initial increase in the output current TFT deviate in the range of 10 ... 40 ampers, and the values of the time interval during which there

83

is such an increase within of 0.2 ... 2 microseconds were modeled.

Simulation results helped to obtain sets of input voltage values for different possible combinations of output parameters TFT and regression models were constructed. The most appropriate and promising seems to be the use of the exponential approximation, since the polynomial approximation of second and third degree when the argument is beyond the considered range 0.2 ... 2 microseconds distorts the real picture of the input voltage variation due to the peculiarities of electromagnetic processes occurring in power supply traction network.

CONCLUSIONS:

1. Thus, it is possible to consider the results of computer simulation obtained on a mathematical model as random variables, and use them in the future as initial data to determine the loss of power.

2. Employing the methods of regression analysis one can replace the original data by the regression equation using the method of least squares.

3. Analysis of the power loss level in the power supply traction network at start can be made on the basis of regression dependences of the current on coordinates of the network cross-section.

4. Power loss in traction network at a starting mode depends on the output parameters TFT (initial step-up of the current and its rise time).

REFERENCES

Pivnyak, G.G. 1990. *Transport with the inductive transmission of energy for coal mines* (in Russian). Moscow: Nedra: 252.

Pivnyak, G.G., Zrazhevskiy, Y.M. & Khovanskaya, E.I. 2004. *Tasks of design of the modes of operations of hauling network of transport with the inductive transmission of energy* (in Russian). Tekhnichna elektrodinamika, Issue 7: 27-30.

Khovanskaya, E.I. & Lysenko, A.G. 2005. *Influence of initial parameters of hauling transformer of frequency on the losses of power in a network* (in Russian). Girnicha elektromehanika ta avtomatika, Isssue 74: 9-13.

Kalantarov, P.L. & Tseytlin, L.A. 1986. *Raschet induktivnostey* (in Russian). Moscow: Energoatomizdat: 488.

Energy Efficiency Improvement of Geotechnical Systems – Pivnyak, Beshta & Alekseyev (eds)
© 2013 Taylor & Francis Group, London, ISBN 978-1-138-00126-8

Estimation of the influence of controllability and observability of models in terms of state variables on the quality of information-measuring system operation

V.I. Korsun & M.A. Doronina
State Higher Educational Institution "National Mining University", Dnipropetrovs'k, Ukraine

ABSTRACT: It is estimated how the properties of controllability and observability of continuous and discrete models in terms of state variables used to describe technical objects and technological processes of mining production, influence the quality of information-measuring systems operation.

1 INTRODUCTION

A progress in the domain of measuring technologies and systems can take place, first of all, due to the development of new measuring methods and instruments and improvement of the existing ones, synthesis of their optimal structures and operation algorithms. Moreover, it is quite useful to employ the achievements of different domains of science, engineering and technology.

It should be noted that nowadays the intellectual and virtual measuring information systems, based on personal computer and modern software, gain in strength and become the part of every day life. And, as a rule, these systems operate in a dynamic mode of measuring different characteristics of the objects being under study.

As to the difficult geotechnical objects, there are integral and nonlinear differential equations being widely used for their description at the present time (including partial derivatives (Glushko 1987; Yalanskiy 2003; Ikonnikova 2007). The attempts to substitute these models for their simplified variants lead to wrong measurement results.

In this case for the qualitative measurement, it is necessary the models of dynamic objects being under study to have such properties as controllability and observability.

2 ESTIMATION OF THE INFLUENCE OF CONTROLLABILITY AND OBSERVABILITY OF CONTINUOUS MODEL IN TERMS OF STATE VARIABLES

According to (Glushko, Yamshchickov &Yalanskiy 1987), controllability is "a property

of the control object consisting in the fact that there are control actions, which can ensure control purpose achievement under conditions of given limitations", and *observability* of a state is "a property of an object consisting in a possibility to evaluate the values of coordinates determining the state of this object in accordance with measured values of coordinates under conditions of given limitations".

Kalman's criteria of controllability and observability are used to evaluate the level of controllability and observability of a linear model in terms of state variables of the n-th order

$$\frac{dx(t)}{dt} = Ax(t) + Bu(t), \quad x(0) = x_0.$$
$$y(t) = Cx(t) + Du(t). \tag{1}$$

with constant matrices A, B and C.

The main point of these criteria consists in the following: for the total controllability and absolute observability of state variables in model (1), it is necessary that

$$rangG = n \text{ and } rangH = n, \tag{2}$$

where

$$G = [B \vdots AB \vdots A^2 B \vdots \cdots \vdots A^{n-1} B], \tag{3}$$

$$H = [C^T \vdots A^T C^T \vdots (A^T)^2 C^T \vdots \cdots \vdots (A^T)^{n-1} C^T]. \tag{4}$$

according to matrices of controllability and observability.

If the matrix is

$$A = diag[\lambda_1, \lambda_2, \ldots, \lambda_n], \tag{5}$$

then for the total controllability and absolute observability of model (1), it is necessary matrix

85

B not to have zero rows and matrix C not to have zero columns.

The model (1) can have in its structure the following submodels (Yalanskiy, Palamarchuk & Rozumnyy 2003):

    a) totally controlled, but absolutely unobserved;

    b) totally controlled and absolutely observed;

    c) totally uncontrolled, but absolutely observed;

    totally uncontrolled and absolutely unobserved

It should be noted that above given division of model (1) into four constituents is possible only if matrix A is a diagonal one.

Only totally controlled and absolutely observed submodel has transfer matrix (b).

$$W(s) = C(sE - A)^{-1} B, \tag{6}$$

For the rest of submodels, it is impossible to create transfer matrices due to lack of either output signal (a), or input signal (c), or input and output signals at the same time (d).

Preserving its structure, the mathematical model (1) can substantially change its properties within parametric variation of matrices B and C.

For example, let the one-dimensional object under study or control be described by means of the mathematical model of the second order (Figure 1):

$$\begin{cases} \dfrac{dx_1(t)}{dt} = 2x_1(t) - 4x_2(t) + b_1 u(t), & x_1(0) = x_{10}. \\ \dfrac{dx_2(t)}{dt} = 7x_1(t) - 9x_2(t) + b_2 u(t), & x_2(0) = x_{20}. \end{cases} \tag{7}$$

$$y(t) = c_1 x_1(t) + c_2 x_2(t).$$

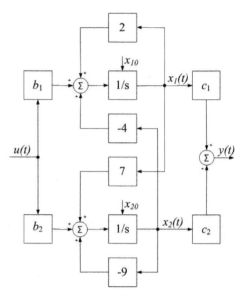

Figure 1. Block diagram of model (7)

Note matrices of controllability G and observability H for model (7):

$$G = [B \vdots AB] = \begin{bmatrix} b_1 & \vdots & 2b_1 - 4b_2 \\ b_2 & \vdots & 7b_1 - 9b_2 \end{bmatrix}, \tag{8}$$

$$H = [C^T \vdots A^T C^T] = \begin{bmatrix} c_1 & \vdots & 2c_1 + 7c_2 \\ c_2 & \vdots & -4c_1 - 9c_2 \end{bmatrix} \tag{9}$$

Determine matrix ranks (8) and (9) by means of finding the maximal order of their minors having nonzero values. Calculate

$$\det G = b_1(7b_1 - 9b_2) - b_2(2b_1 - 4b_2) = \\ = (b_1 - b_2)(7b_1 - 4b_2) \tag{10}$$

and

$$\det H = c_1(-4c_1 - 9c_2) - c_2(2c_1 + 7c_2) = \\ = -(4c_1 + 7c_2)(c_1 + c_2) \tag{11}$$

It results from (10) and (11), if parameters b_1 and b_2 satisfy the inequality system

$$\begin{cases} b_1 - b_2 \neq 0 \\ 7b_1 - 4b_2 \neq 0 \end{cases} \tag{12}$$

and parameters c_1 and c_2 satisfy the inequality system

$$\begin{cases} 4c_1 + 7c_2 \neq 0 \\ c_1 + c_2 \neq 0 \end{cases}, \tag{13}$$

(the above-mentioned corresponds to the totally controlled and absolutely observed model(1)), then the transfer function being calculated in accordance with expression (13) is as follows:

86

$$W(s) = \frac{k_1 s + k_2}{(s+5)(s+2)}, \qquad (14)$$

where $k_1 = b_1 c_1 + b_2 c_2$ and

$k_2 = 9b_1 c_1 - 4b_2 c_1 + 7b_1 c_2 - 2b_2 c_2$

But if $b_1 - b_2 = 0$ and c_1 and c_2 still satisfy the system (13), then assuming $b_1 = b_2 = b$, obtain the transfer function of the first order

$$W(s) = \frac{b(c_1 + c_2)}{s+2} \qquad (15)$$

which corresponds to the mathematical model in terms of state variables of the second order (Figure 2).

$$\begin{cases} \dfrac{dx_1(t)}{dt} = 2x_1(t) - 4x_2(t) + bu(t), & x_1(0) = x_{10}. \\ \dfrac{dx_2(t)}{dt} = 7x_1(t) - 9x_2(t) + bu(t), & x_2(0) = x_{20}. \end{cases} \qquad (16)$$

$y(t) = c_1 x_1(t) + c_2 x_2(t)$

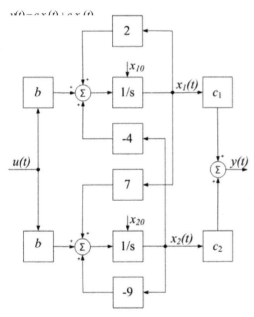

Figure 2. Block diagram of model (16)

Here, variables $x_1(t)$ and $x_2(t)$ at different values $u(t)$ get their values on the plane (x_1, x_2) in a band, which is limited from the one side by a line passing through the origin of coordinates along its own vector $[b \quad b]^T$, and from the other side by a line being drawn through a point (x_{10}, x_{20}) parallel to the indicated eigenvector.

If $c_1 = c$ and $c_2 = -c$ are equated in model (16) (to ensure the equality of $c_1 + c_2 = 0$), obtain the following model (Figure 3)

$$\begin{cases} \dfrac{dx_1(t)}{dt} = 2x_1(t) - 4x_2(t) + bu(t), & x_1(0) = x_{10}. \\ \dfrac{dx_2(t)}{dt} = 7x_1(t) - 9x_2(t) + bu(t), & x_2(0) = x_{20}. \end{cases} \qquad (17)$$

$y(t) = c_1 x_1(t) - c_2 x_2(t)$.

There is no cause-and-effect relation between its input $u(t)$ and output $y(t)$ signals. Here is: $W(s) = 0$.

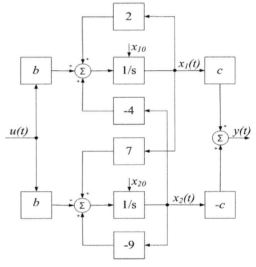

Figure 3. Block diagram of model (17)

This fact can be easily explained if model (1) is transformed by means of substitution $[x_1(t) \quad x_2(t)]^T = R[z_1(t) \quad z_2(t)]^T$, where $R = [u_1 \vdots u_2]$ is modal matrix.

Here $u_1 = \begin{bmatrix} \dfrac{4}{7} & 1 \end{bmatrix}^T$ and $u_2 = \begin{bmatrix} 1 & 1 \end{bmatrix}^T$ are eigenvectors of matrix

$$A = \begin{bmatrix} 2 & -4 \\ 7 & -9 \end{bmatrix}, \qquad (18)$$

of model (1).

The transformed model is as follows (Figure 4):

$$\begin{cases} \dfrac{dz_1(t)}{dt} = -5z_1(t) - \dfrac{7}{3}(b_1 - b_2)u(t), & z_1(0) = \dfrac{7(x_{20} - x_{10})}{3}. \\ \dfrac{dz_2(t)}{dt} = -2z_2(t) + bu(t), & z_2(0) = \dfrac{7x_{10} - 4x_{20}}{3}. \end{cases} \qquad (19)$$

87

$$y(t) = \frac{4c_1 + 7c_2}{7} z_1(t) + (c_1 + c_2)z_2(t).$$

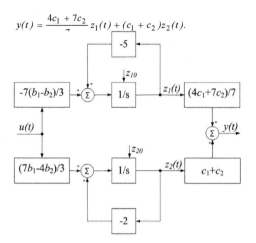

Figure 4. Block diagram of model (19)

Assume $b_1 = b_2 = b$, then system (19) is as follows (Figure 5):

$$\begin{cases} \dfrac{dz_1(t)}{dt} = -5z_1(t), & z_1(0) = \dfrac{7(x_{20} - x_{10})}{3} \\ \dfrac{dz_2(t)}{dt} = -2z_2(t) + bu(t), & z_2(0) = \dfrac{7x_{10} - 4x_{20}}{3} . \end{cases} \quad (20)$$

$$y(t) = \frac{4c_1 + 7c_2}{7} z_1(t) + (c_1 + c_2)z_2(t).$$

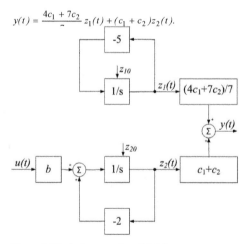

Figure 5. Block diagram of model (20)

It is obvious that $z_1(t)$ is uncontrolled and isn't involved in the connection formation between $y(t)$ and $u(t)$. The transfer function (20) of the first order is formed both by means of the second differential equation of system (15) and the output equation.

But if in addition to correlation $b_1 = b_2 = b$, the correlation $c_1 + c_2 = 0$ is also observed (that

can be if $c_1 = c$ and $c_2 = -c$), obtain the model (Figure 6)

$$\begin{cases} \dfrac{dz_1(t)}{dt} = -5z_1(t), & z_1(0) = \dfrac{7(x_{20} - x_{10})}{3} . \\ \dfrac{dz_2(t)}{dt} = -2z_2(t) + bu(t), & z_2(0) = \dfrac{7x_{10} - 4x_{20}}{3} . \end{cases} \quad (21)$$

$$y(t) = -\frac{3}{7} c z_1(t) .$$

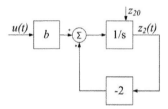

Figure 6. Block diagram of model (21)

Analyzing model (21), resulted from model transformation (17), one can see that control action $u(t)$ changes only variable $z_2(t)$, and output signal $y(t)$ is formed by variable $z_1(t)$, which does not depend on $z_2(t)$. Thus, as it was pointed out above, there is no cause-and-result relation between signals $u(t)$ and $y(t)$ of models (21) and (17).

In case of the mathematical model, this statement is obvious (21), but in model (17) the lack of dependence between input $u(t)$ and output $y(t)$ signals can be explained in the following way. After applying Laplace transformation to the differential equation system (17), it is as follows

$$\begin{cases} sX_1(s) - x_{10} - 2X_1(s) - 4X_2(s) + bU(s) \\ sX_2(s) - x_{20} - 7X_1(s) - 9X_2(s) + bU(s) \end{cases} \quad (22)$$

The system is solved (21):

$$X_1(s) = \frac{(s+9)(x_{10} + bU(s)) - 4(x_{20} + bU(s))}{s^2 + 7s + 10}, \quad (23)$$

$$X_2(s) = \frac{(s-2)(x_{20} + bU(s)) + 7(x_{10} + bU(s))}{s^2 + 7s + 10} . \quad (24)$$

88

Substitute obtained values $X_1(s)$ and $X_2(s)$ into expression $Y(s) = cX_1(s) - cX_2(s)$.

Obtain $Y(s) = \dfrac{c(x_{10} - x_{20})}{s + 5}$, or in the temporary realm:

$$y(t) = c(x_{10} - x_{20})e^{-5t}. \tag{25}$$

The last expression testifies that output signal of object $y(t)$ doesn't depend on its input signal $u(t)$. Coincidence with the conclusions concerning model (17) is observed.

In other words, if there are all connections in block diagram (Figure 3) of model (17), its output signal is always described by means of expression (25) at any input signal $u(t)$.

3 ESTIMATION OF THE INFLUENCE OF CONTROLLABILITY AND OBSERVABILITY OF DISCRETE MODEL IN TERMS OF STATE VARIABLES

Similar conclusions can also be drawn as far as discrete model is concerned, for example

$$\begin{cases} \Delta x_1[kT] = -0.8x_1[kT] + 0.6x_2[kT] + b_1u[kT], \\ \Delta x_2[kT] = 0.7x_1[kT] - 0.7x_2[kT] + b_2u[kT], \end{cases}$$

$$\begin{aligned} x_1[0] &= x_{10} \\ x_2[0] &= x_{20} \end{aligned} \tag{26}$$

$$y[kT] = c_1x_1[kT] + c_2x_2[kT].$$

For this purpose, substitute $\Delta x_1[kT]$ and $\Delta x_2[kT]$ for their values $x_1[(k+1)T] - x_1[kT]$ and $x_2[(k+1)T] - x_2[kT]$ respectively in system (26). Obtain:

$$\begin{cases} x_1[(k+1)T] = 0.2x_1[kT] + 0.6x_2[kT] + b_1u[kT], \\ x_2[(k+1)T] = 0.7x_1[kT] + 0.3x_2[kT] + b_2u[kT], \end{cases}$$

$$\begin{aligned} x_1[0] &= x_{10} \\ x_2[0] &= x_{20} \end{aligned} \tag{27}$$

$$y[kT] = c_1x_1[kT] + c_2x_2[kT].$$

It is possible to apply the same transformations used for model (7) to model (27). As a result, obtain the models with different discrete transfer functions (including transfer function $W[z] = 0$ as in a case of model (17) and its analogue (19)).

Thus, continuous models as well as discrete ones in terms of state variables, at which parameter and characteristic estimation the information and measurement systems are aimed, having a whole range of dependences between individual elements, don't guarantee the cause-and-effect relation between input and output signals.

4 DEPENDENCE OF CONTROLLABILITY AND OBSERVABILITY LEVELS ON TRANSITION FROM CONTINUOUS MODEL TO ITS DISCRETE ANALOGUE

It should be noted, that the transition from continuous model to its discrete analogue leads to decrease of controllability and observability level of appropriate state variables at some periods of time quantization.

Let's demonstrate it solving matrix differential equation (1) under initial condition $x(t_0) = x_0$:

$$x(t) = e^{A(t-t_0)}x(t_0) + \int_{t_0}^{t} e^{A(t-\tau)}Bu(\tau)d\tau. \tag{28}$$

Being given $t_0 = kT$ and having assumed $u(t) = u[kT]$ under condition $kT \le t < (k+1)T$, obtain

$$x(t) = e^{A(t-kT)}x[kT] + \int_{kT}^{t} e^{A(t-\tau)}Bd\tau \cdot u[kT]. \tag{29}$$

At $t = (k+1)T$, the discrete analogue of differential equation (1) is as follows:

$$\begin{aligned} x[(k+1)T] &= A_d x[kT] + B_d u[kT], \\ x[0] &= x_0. \end{aligned} \tag{30}$$

$$y[kT] = C_d x[kT].$$

where

$$A_d = e^{AT}, \quad B_d = \int_0^T e^{Az}Bzdz, \quad C_d = C. \tag{31}$$

Let's demonstrate a lack of controllability and observability of discrete model which has been obtained on the basis of expressions (30) from totally controlled and observed continuous model

$$\frac{d^2y(t)}{dt^2} + 2\frac{dy(t)}{dt} + 101y(t) = 1 \cdot u(t), \quad y(0) = 0,$$

$$\frac{dy(0)}{dt} = 0. \tag{32}$$

Note the differential equation (32) in the form of a model in terms of state variables (1):

$$\begin{cases} \dfrac{dx_1(t)}{dt} = x_2(t), \\ \dfrac{dx_2(t)}{dt} = -101x_1(t) - 2x_2(t) + 1 \cdot u(t), \end{cases}$$

$$x_1(0) = 0.$$
$$x_2(0) = 0. \tag{33}$$
$$y(t) = x_1(t)$$

Here matrices A, B and C are respectively equal

$$A = \begin{bmatrix} 0 & 1 \\ -101 & -2 \end{bmatrix}, \quad B = \begin{bmatrix} 0 \\ 1 \end{bmatrix}, \quad C = \begin{bmatrix} 1 & 0 \end{bmatrix}.$$

For system (33) from the given matrix A, obtain the fundamental matrix

$$\Phi(t) = \begin{bmatrix} \phi_{11}(t) & \phi_{12}(t) \\ \phi_{21}(t) & \phi_{22}(t) \end{bmatrix}, \tag{34}$$

Where

$$\phi_{11}(t) = e^{-1}(cos(10t) + 0.1\,sin(10t)),$$
$$\phi_{12}(t) = 0.1e^{-1}\,sin(10t),$$
$$\phi_{21}(t) = -10.1e^{-1}\,sin(10t),$$
$$\phi_{22}(t) = e^{-1}(cos(10t) - 0.1\,sin(10t)). \tag{35}$$

Having used the values of fundamental matrix $\Phi(t)$, of controllability matrix B and observability matrix C of continuous system (33), obtain

$$A_d = \begin{bmatrix} e^{-T}(cos(10T) + 0.1\sin(10T)) & 0.1e^{-T}\sin(10T) \\ -10.1e^{-T}\sin(10T) & e^{-T}(cos(10T) - 0.1\sin(10T)) \end{bmatrix} \tag{36}$$

$$B_d = \begin{bmatrix} \dfrac{-e^{-T}(10\cos(10T) + \sin(10T))}{1010} \\ 0.1e^{-T}\sin(10T) \end{bmatrix} \tag{37}$$

$$C^d = \begin{bmatrix} 1 & 0 \end{bmatrix}. \tag{38}$$

Since the order of the given model is $n=2$, then its matrix of controllability is as follows:

$$G = \begin{bmatrix} B_d & \vdots & A_d B_d \end{bmatrix}. \tag{39}$$

Having substituted the values of matrixes A_d and B_d into matrix G, obtain

$$G = \begin{bmatrix} \dfrac{-e^{-T}(10\cos(10T) + \sin(10T))}{1010} & \dfrac{-e^{-2T}(10\cos^2(10T) + \cos(10T)\sin(10T) - 5)}{505} \\ 0.1e^{-T}\sin(10T) & \dfrac{e^{-2T}\cos(10T)\sin(10T)}{5} \end{bmatrix} \tag{40}$$

Calculate

$$det\,G = \frac{-e^{-3T}\,sin(10T)}{1010}. \tag{41}$$

It results from expression (41), that at $10T = k\pi$, $(k = 1,2,3...)$ $det(G) = 0$. And discrete analogue of continuous model (32) (and (33)) is partially controlled.

Herewith, matrix of controllability of discrete model

$$G = \begin{bmatrix} \dfrac{-e^{-T}cos(k\pi)}{101} & \dfrac{-e^{-2T}(2\cos^2(k\pi) - 1)}{101} \\ 0 & 0 \end{bmatrix}, \tag{42}$$

has $rang(G) = 1$.

In its turn, matrix of observability of discrete model (30), (36),(37),(38) is as follows:

$$H = \begin{bmatrix} C_d^T & \vdots & A_d^T C_d^T \end{bmatrix}. \tag{43}$$

Having substituted the values of matrices A_d and C_d into matrix H, obtain

$$H = \begin{bmatrix} 1 & e^{-T}\,cos(k\pi) \\ 0 & 0 \end{bmatrix}. \tag{44}$$

Here also, at $10T = k\pi$, $(k = 1,2,3...)$ $det(H) = 0$, and discrete analogue of the initial continuous model is observed to some extent.

Under these conditions, matrix of observability of discrete model

$$H = \begin{bmatrix} 1 & e^{-T}\,cos(k\pi) \\ 0 & 0 \end{bmatrix}, \tag{45}$$

has $rang(H) = 1$.

5 CONCLUSIONS

Above-mentioned situations reveal the cases when cause-and-effect relation between input and output signals of models is not observed, or it has partial or complete admission. Moreover, continuous models as well as discrete ones have the same structure.

It is also demonstrated that within transition from continuous model to its discrete representation, the decrease of controllability and observability level of the model is possible. It inevitably has an impact on performance quality of information and measurement systems, which are intended for operation with such models.

This fact should be taken into account to estimate dynamic model parameters with due regard to the values of their input and output signals.

REFERENCES

Glushko, V.T., Yamshchickov, V.S., Yalanskiy, A.A. 1987. *Geogphysical Control in Mine Tonnels* (in Russian). Moscow: Nedra: 278.

Yalanskiy, A.A., Palamarchuk, T.A., Rozumnyy, S.N. 2003. *Peculiarities and diagnostics of self-organization processes in rock massif in situ mining* (in Russian). MGGU: Gornyy informatsyonno-analiticheskiy bulleten, Issue 3: 151-154.

Ikonnikova, N.A. 2007. *Peculiarities of modelling the chaotic process dynamics in determined systems by the methods of analytical mechanics* (in Russian). Geoteckhnicheskaya mekhanika, Issue 73: 263-280.

Krasovskii, A.A. 1987. *Handbook on Automatic Control Theory* (in Russian). Moskow: Ed. Nauka: 711.

Energy Efficiency Improvement of Geotechnical Systems – Pivnyak, Beshta & Alekseyev (eds)
© 2013 Taylor & Francis Group, London, ISBN 978-1-138-00126-8

A solution to the problem of frequency compatibility between drive systems and dynamic parameters of drilling rigs

V. Khilov
State Higher Educational Institution "National Mining University", Dnipropetrovs'k, Ukraine

ABSTRACT: The theory of the functioning of quick-acting automated electric drives of drilling rigs has been proposed. The frequency peculiarities of the control objects of drilling rig drives have been investigated. The task of dynamic compatibility between the drive frequency characteristics and the characteristics of multimass mechanisms, lumped distributed parameters and variable attached masses has been solved. The control algorithms, depressing the fluctuations in the control system, have been presented. The ideas of neuro-fuzzy control for the depression of fluctuations in the control system has been substantiated.

1 INTRODUCTION

The common industrial drives with active correction and two-loops systems of control through the speed channels of rotation and excitation of both DC and AC motors are a technical standard of drive systems with increased requirements to static and dynamic control modes (Quang 2010; Novotny 1996; Boldea 1992).

The control technology of complicated dynamic objects on the basis of automated electric DC and AC drives with the consecutive correction of characteristics occupies an important place in the modern approach to the optimization of dynamic work modes of machines and installations (Boldea 2006; Khilov 2004; Pivnyak 2005). The characteristic feature of such systems is the presence of a thyristor rectifier for a DC drive and a voltage transistor autonomous inverter with the amplitude-pulse modulation for an AC drive, as well as special transmissions which transmit torque from the engine to the executive device. At the same time, masses and elasticities of separate units of the mechanical system can be lumped and distributed (drilling rigs, excavators, band conveyors, mine winder installations etc (Khilov & Plakhotnik 2004; Pivnyak 2003). Such control objects have low natural frequencies. The DC drive systems, due to the inertia of a thyristor converter and mechanical inertia of motor armature are low frequency filters of good quality having a bandwidth of up to 50 rad/s. The AC drive systems, due to lower mechanical inertia of a

motor and higher switching frequency of a pulse-impulse electrical modulator, have a bandwidth of up to 200 rad/s. Therefore, the modernization of the electric drive with the replacement of a DC system by an AC one, greatly changes the work dynamics of the electromechanical installation (Khilov 2006).

2 FORMULATION THE PROBLEM

Using the method of active consecutive correction and ideas of fuzzy control, let us find additional contour corrective controllers which depress the influence of transmission elastic properties on the control system dynamics.

3 MATERIALS FOR RESEARCH

Additional dynamic units, conditioned by the presence of transmission elastic properties, are described by the fractional-rational functions of higher degrees (Pivnyak & Beshta 2005, Khilov 2006). The characteristic frequency values of an additional transfer function are less than the corresponding control loop cutoff frequency, Fig.1. A generalized structural scheme of the control object is presented in the form of transfer functions of the object control compensated part W_{okn} (s) and transfer functions of the object control uncompensated part, conditioned by the transmission elastic properties and this structural scheme is given as a fractionally rational function $\Sigma a_{mn}/\Sigma b_{mn}$.

In the studied drive systems, the control loop of each parameter (coordinate) contains, as a rule, not only one "big" time constant, compensated by the action of a controller, but also a fractional-rational function with time constants fully comparable with the singled out "big" time constant, that greatly complicates the system synthesis and does not make it possible to apply the control algorithms used in common industrial drives.

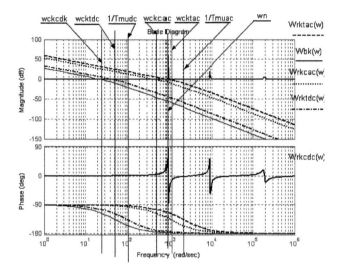

Figure 1. Logarithmic amplitude and phase frequency characteristics of open-loop current and rotating frequency of DC drives $W_{rktdc}(w)$, $W_{rkcdc}(w)$, AC drives $W_{rktac}(w)$, $W_{rkcac}(w)$ and a drilling tube $W_{bk}(w)$ ($T_{\mu ac}$, $T_{\mu dc}$ – "small" uncompensated time constants of AC and DC drives, the cutoff frequency of current loops and rotating frequency of DC drives W_{cktdc}, W_{ckcdc} and AC drives W_{cktac}, W_{ckcac})

The control object is structurally broken up into two units: compensated and uncompensated by a classic controller parts. The compensated part is chosen in the form of a dynamic unit not exceeding the second order with one explicitly expressed "big" time constant. Then the processes in the loop are effectively corrected by a classic controller with the PID (proportional-integral-differential) characteristic or its derivatives (PI – proportional-integral, PD – proportional-differential, P – proportional, I – integral).

The sought-for control laws of the investigated electric drives are determined by technological specificities with the account of restrictions imposed on the parameters of electrical and mechanical parts of the drive system. Restrictions are imposed on the parameters of the mechanical energy flow, magnitudes of accelerations and jerks permitted by the transmission, the overload capacity of the engine, its permissible heating, etc. These factors combined with the type of perturbation determine allowable laws of change both of the external (output) and all the intermediate (internal) coordinates of the drive system. Determination of a classical algorithms of the functioning of controllers is conducted by a sequential correction of dynamic characteristics of each loop taken separately, beginning with the most fast-acting – the internal $m = 1$, up to the outer loop $m = n$ with a minimum speed. The term 'optimization' should be understood as the adduction of dynamic parameters of the closed loop in accordance with the limitations imposed on the quality of the transition process.

The generalized transfer function of the optimized loop object under the made assumptions may be written as

$$W_{on}(p) = W_{okn}(p) \cdot W_{onn(p)} =$$
$$\frac{2 \cdot \xi'_n \cdot T'_n \cdot s + 1}{T_n^2 \cdot s^2 + 2 \cdot \xi_n \cdot T_n \cdot s + 1} \cdot \frac{\sum a_{mn} \cdot s^m}{\sum b_{mn} \cdot s^m}, \quad (1)$$

where T_n, T_n^I – time constants, ξ_n, ξ_n^I – damping coefficients.

Here the dynamic unit $W_{okn}(s)$ corresponds to the units the effect of which is compensated by a classic controller. The second dynamic unit is presented in the form of fractional and rational expressions, the actions of which, due to the complicated character of the compensation algorithm, cannot be depressed by a classical PID controller. If to attribute the time constants of fractional and rational transfer functions to the "small" uncompensated time constant of the control loop, its fast-action will significantly decrease. The latter circumstance would adversely affect the fast-action of the entire control system. The allowable limit of the loop speed is determined by the choice of a "small" uncompensated time constant of the optimized control loop.

To depress elastic vibrations of transmission it is possible to include additional control units parallel to the main controllers (distributed additional classic controllers) or to recalculate the action of parallel additional controllers on the input of the semiconductor converter – the output of the current controller (concentrated additional classical controller). The latter is preferable, as the corrective action is fed to the fastest loop. Under such mode of depressing the fluctuation processes, the algorithms of controllers' work in the control system are determined as follows:

$$W_{n(s)} = \Sigma b_{mn} s^m \ \Sigma a_{mn} s^m. \tag{2}$$

For the compensation of additional transfer functions in series with a classic controller, the additional controller with a transfer function is switched on.

$$W_n''(p) = \frac{1}{p \cdot \sigma_n \cdot W_{okn}(p)} \cdot \frac{\sum (b_{mn} - a_{mn}) \cdot p^m}{\sum a_{nm} \cdot p^m} \times$$
$$\times \prod_{k=n-1}^{1} \left(\frac{1}{p \cdot \sigma_k \cdot W_{okk}(p)} \cdot \frac{\sum b_{nk} \cdot p^n}{\sum a_{nk} \cdot p^n} \right)$$

If to transfer the actions of all additional controllers to the output of the internal controller, then instead of n included parallel to each controller it is possible to find one controller, which embraces parallel classic controllers and replaces the action of these controllers. But in this case the algorithm of one controller is becoming more complicated than the algorithm of each controller separately Tab. 1.

$$W_n''(s) = \frac{1}{s \cdot \sigma_n \cdot W_{okn}(s)} \cdot \frac{\sum (b_{mn} - a_{mn}) \cdot s^m}{\sum a_{nm} \cdot s^m} \times \tag{3}$$
$$\times \prod_{k=n-1}^{1} \left(\frac{1}{s \cdot \sigma_k \cdot W_{okk}(s)} \cdot \frac{\sum b_{nk} \cdot s^n}{\sum a_{nk} \cdot s^n} \right)$$

Table 1. The loop controllers algorithms with the account of dynamic properties of a control object

The loop controlled value	The loop controllers algorithm	Correction algorithm of influence additional units in the loop
Stator current active component	$\dfrac{T_2^2 + 2\xi_2 T_2 s + 1}{a_T T_\mu^2 s^2 T_{em}}$ $\gamma \dfrac{K_p}{R_s + K_r^2 R_r}$	$\dfrac{\left(T_4^2 - T_E^2\right)s^2 + 2(\xi_4 - \xi_E)T_4 s}{T_E^2 s^2 + 2\xi_E T_E s + 1}$ $\dfrac{T_2^2 s^2 + 2\xi_2 T_2 s + 1}{a_T T_\mu s K_p / \left(R_s + K_r^2 R_r\right)}$
The motor frequency rotation	$\dfrac{\gamma T_{m1}}{a_c a_T T_\mu} \dfrac{K_r p_n \psi_r}{R_s + K_r^2 R_r}$ $\dfrac{s a_c a_T T_\mu + 1}{s a_c a_T T_\mu}$	$\dfrac{(\gamma - 1)s^2}{T_E^2 \gamma s^2 + 2\xi_E T_E s + 1} \dfrac{\gamma T_{m1}}{a_c a_T T_\mu}$ $\dfrac{K_r p_n \psi_r}{R_s + K_r^2 R_r} \dfrac{s b_c a_c T_\mu + 1}{s b_c a_c T_\mu}$

The loop controlled value	The loop controllers algorithm	Correction algorithm of influence additional units in the loop
The tube frequency rotation	$\dfrac{\gamma T_{m1}}{a_c a_T T_\mu}\; \dfrac{K_r p_n \psi_r}{R_s + K_r^2 R_r}\; \dfrac{s a_c a_T T_\mu + 1}{s a_c a_T T_\mu}$	$\dfrac{\gamma T_E^2 s^2}{2\xi_E T_E s + 1}\; \dfrac{\gamma T_{m1}}{a_c a_T T_\mu}$ $\dfrac{K_r p_n \psi_r}{R_s + K_r^2 R_r}\; \dfrac{s b_c a_c T_\mu + 1}{s b_c a_c T_\mu}$

where K_P, T_μ – the semiconductor converter transfer coefficient and the "small non-compensable" time constant; R_S, R_r – the resistive resistance of stator and rotor windings; T_{M1}, T_{M2} – the mechanical time constants of a electric motor; γ – the ratio coefficient of inertial masses of the system; K_r, p_n – the adduced factor and the number of pole pairs of the induction motor; ψ_r – the rotor flux linkage; ξ_E, T_E – the damping factor and the time constant of elastic vibrations of a tube; a_C – the speed loop tuning; a_T – the current loop tuning coefficient.

All correction algorithms of dynamic processes contain derivatives which are not lower than the second-degree of the error signal, therefore if noises are present in the signal they will affect the efficiency of such compensation. In addition, changing the parameters of control objects results in the instability of the characteristic frequencies of additional dynamic elements, so adjustment of correction algorithms is possible only in one operating point of loops. Changing the number of added masses results not in the compensation of elastic vibrations by additional correction units, but in the increased vibration of the control system. Therefore, the correction algorithms should have the capacity of adapting to the variable parameters of the control object.

In connection with the above mentioned, we solve the problem of elastic vibrations compensation in the regulation systems with consecutive correction by applying the principles of fuzzy control for complicated objects, which also include electric drives, under the presence of elastic constraints in the transmission. This makes it possible to preserve well-known advantages of systems with the consecutive correction of dynamic parameters (Pivnyak 2004; Khilov 2011; Khilov &

Glukhova 2004; Terekhov 2001; Terekhov 1996).

The approach, based on the theory of fuzzy sets, has its distinctive characteristic features: in addition to the numeric variables the fuzzy values, so-called "linguistic" variables, are used; simple relations between the variables are described with the help of fuzzy utterances; complicated relations are described with the help of fuzzy algorithms (Zadeh 1976). This approach gives approximate, but at the same time effective ways of describing the behavior of the systems which are so complicated that for the correction of their dynamics, adaptive controllers with polynomials of high degrees are required, both in the numerator and in the denominator.

To preserve the benefits of the control systems with active consecutive correction, parallel to the classic controllers we use a fuzzy controller, the functions of which consists in the depression of vibrations arising in the control system due to the elastic vibrations in the transmission (Fig. 2).

To depress the elastic vibrations of transmission in the control loops we take into consideration not only a signal proportional to the error between specified and actual values of a controlled magnitude but also a rate of change and the integral of the marked error, which corresponds to the fuzzy controller with the dynamic PID characteristic. The structure of a fuzzy controller foresees the selection of input functions of possession, rules of processing the terms, search of the output signal.

The synthesis of a fuzzy controller is rationally made in two stages. At the first stage, we make a choice of the number of membership functions on the basis of the cluster analysis of a fuzzy controller input signal vector. At the second stage, based on the algorithms for adaptive neural networks, we formulate the rules of processing   the terms.

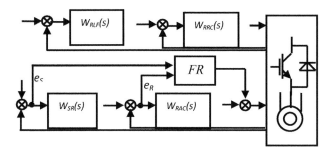

Figure 2. The functional scheme of the fuzzy control with additional correction, where ($W_{RLF}(s)$ – the flux linkage controller; $W_{RRC}(s)$, $W_{RAC}(s)$ – the reactive and active current controllers; $W_{SR}(s)$, FR – the speed and fuzzy controllers; e_{SR}, e_{RC} – the speed and current errors)

The process of an abrupt increase of a motor speed from zero to the nominal level, of a reverse speed from the nominal positive level till the nominal negative one, and then back from the negative to the positive nominal one, allows investigators to create the most characteristic data selection of variable signals at the input and output of a fuzzy controller, used for the training the membership functions and rules for processing of terms. Sampling signal values at the input and output of a fuzzy controller is supplemented by signals under the same operating regimes but with an abrupt change of the moment of resistance on the motor shaft.

The initial information for clustering is an observation matrix of input signals of the fuzzy controller D, which is formed according to the oscillogram data curves of the signal waveform at the input of a fuzzy controller where $E(t)$– the error at the controller input.

$$D = \begin{bmatrix} E(t_1) & dE/dt|_{t1} & \int E \cdot dt|_{t1} \\ E(t_2) & dE/dt|_{t2} & \int E \cdot dt|_{t2} \\ \ldots & \ldots & \ldots \\ E(t_n) & dE/dt|_{tm} & \int E \cdot dt|_{tn} \end{bmatrix}, \quad (4)$$

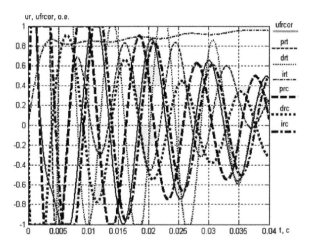

Figure 3. Error oscillograms at the inputs of the controllers: current, speed, rate of change and the integral of the error, where $E_P(p)$– the error on the input controllers; $RE_P(t)$– the rate of change of the error at the controllers input; $IE_{PT}(p)$ – the integral component of the error in the depression of vibrations in the current loop and speed.

These signals can be calculated analytically, using the found dynamic model of the control loop, Fig. 3.

Each row of the matrix D represents itself as the values of the three features of one of the objects of clustering for each controller, respectively proportional, differential and the integral parts. The task of clustering is to divide the objects located in the matrix D into several subsets (clusters) in which objects are more similar to each other than to the objects of other clusters. In a metric space "similarity" is determined by the distance. The distance is calculated both between initial object (with the rows of matrix D) and from these objects to the clusters prototype. In this case, the coordinates of the prototypes are unknown – they are found simultaneously with the division of the data into clusters.

In the task being solved, any a priori assumptions concerning the number of fuzzy clusters are absent, therefore to determine the number of clusters we use a subtractive clustering method offered by R. Yager and D. Filev (Yager & Filev 1984). The idea of this method is that each point of data is accepted as a center of a potential cluster, after which the ability of any point to represent the cluster center is computed. This quantitative measure is based on the estimate of density of data points near the corresponding center of the cluster.

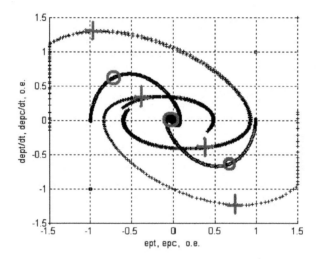

Figure 4. The result of the cluster analysis of the output signal of current and speed fuzzy controllers from the error and rate of error at the inputs

The vector values of the fuzzy controller input signal were computed according to the algorithm of fuzzy clustering. The calculation result is shown in Fig. 4. The found centers of clustering are marked by large crosses in Fig. 4 (four cluster centers for the current controller) and large circles (three cluster centers for the speed controller). The number of cluster centers determines the numbers of membership functions, that is, four membership functions are needed for the current controller and three – for the speed controller.

Having determined the number of clusters for the input variables, we move to the second stage of the fuzzy controller synthesis. For this purpose we use the possibilities of training artificial neural networks in order to find the output membership functions and the relative position of both input and output membership functions.

To find the rule base according to the initial membership functions we use the apparatus of hybrid networks in which the conclusions are made on the basis of fuzzy logic, but the corresponding membership functions are tuned on the basis of the algorithms of neural networks, the so-called adaptive neuro-fuzzy inference system (ANFIS) (Leonenkov 2003). Such training systems use multiples of input and output values and they distribute the membership functions according to the data vector D. The parameters of membership functions change in accordance with the signals presented for training. Calculation of these parameters is done in accordance with the calculated gradient vector, which controls the deviation of the initial coordinate of the neural network from the specified value in a set of input / output data provided for training. D. The backpropagation algorithm (Medvedev 2002) as well as the combination of the least squares method and the backpropagation algorithm have been used for training. These algorithms allow the neuro-fuzzy controller to be effectively taught from the data set at the input and output.

A fuzzy neural network consists of four layers of neurons:

the output nodes of the first layer are the numerical values of the membership functions in accordance with the values of the input signals;

the degrees of truthfulness of prerequisites of the formed rule-based knowledge system are the neuron outputs of the second layer and they are calculated in accordance with the expressions (knowledge base);

the neurons of the third layer calculate the values of the output membership functions;

the single neuron of the fourth layer computes the network output.

Training of the neuro-fuzzy network of the ANFIS type was done under the following conditions:

the number of epochs accepted for training is 100;

the number of input membership functions – N;

the training algorithms: of back-propagation and a hybrid one (Medvedev 2002);

the checked membership functions: triangular, trapezoidal, generalized bell-shaped, Gaussian, double Gaussian, π-function, double sigmoidal, the product of two sigmoidal functions.

the accepted output functions: of the Suqeno, const or line orders.

Proceeding from the obtained data for the errors of training, there is no need to take more complicated input membership functions than the triangular or Gaussian ones.

To define the membership functions the typical Gaussian function was used

$$gaussmf(x, \sigma, c) = e^{-\left(\frac{x-c}{\sigma}\right)}, \qquad (5)$$

where c – the displacement; σ – a width of the Gaussian function, Tab. 2.

Table 2. The Gaussian membership functions parameters

The fuzzy controller input signal	The Gaussian membership functions parameters σ / c			
	$inmf1$	$inmf2$	$inmf3$	$inmf4$
e_{PT}	0.3/-1.4	0.1/-04	0.2/0.2	0.4/1.4
de_{PT}/dt	0.6/-1.3	0.8/-0.5	0.2/0.7	0.4/1.5
e_{PC}	0.4/-0.5	0.5/0.2	0.4/0.9	–
de_{PC}/dt	0.5/-0.9	0.6/0.06	0.5/0.9	–

In the algorithms of the current and speed fuzzy controllers it is reasonable to use the two components – proportional and differential. It is possible to use the output membership functions of the Sugeno type and of the zero-order.

The rule base, obtained on the fuzzy-neural network is shown in tab. 3. The following denotations are accepted in the table: $\overline{C21, C216}$ – the numerical values of the initial membership functions of the Sugeno type and of the zero order for the current controller and $\overline{C11, C19}$ – for the speed controller.

From the comparison of the signals corresponding to the reality at the output of the fuzzy controller and the one to be reproduced it

follows that in the periods of occurrence of free transition components (change of a specified signal at the loop inputs), the output signal of the fuzzy controller corresponds to the desired one only by sign and not by value. In the moments of changing the specified signal, the loop dynamic is mainly formed by the classic controllers of active current and speed with the depression of circulating electromotive force (EMF). Under the transition to the forced values of controlled variables, real and desired output voltages of the PD fuzzy controller coincide. Thus, in the forced mode the fuzzy controller will significantly affect the dynamic processes in the control loops.

An analysis of transition processes shows that inclusion of a fuzzy controller parallel to a classical one results not only in an increase of a loop speed, but also in an increase of deregulation as compared with the modular optimum tuning. In addition to the increased deregulation in the time of the second matching the increased oscillation is observed. To eliminate these adverse effects, the current loop tuning coefficient increases from the value of $a_T=2$, which should be for the modular optimum to the value of $a_T=3$. This decreases the control system speed practically to the value of the modular optimum tuning. Simultaneously, the deregulation and loop speed oscillation significantly decrease to the values less than at modular optimum tuning.

Table 3. Fuzzy controller settings

		Error at the input of the current e_{PT}/ speed e_{PC} controller			
		in1 mf1/ in1 mf2	*in1 mf2/ in1 mf3*	*in1 mf3/ in1 mf1*	*in1 mf4*
Rate of current error change de_{PT}/dt and speed error change de_{PC}/dt	*in2 mf1/ in2 mf1*	*C21=0.6504/ C11=14.03*	*C22=1.515/ C12=-0.5238*	*C23=0.9749/ C13=-7.919*	*C24= 0.559*
	in2 mf2/ in2 mf2	*C25=5.225/ C14=-4.081*	*C26=- 1.565/C15= -6.022*	*C27= 3.614/C16= 5.441*	*C28=- 0.00256*
	in2 mf3/ in2 mf3	*C29=-0.906/ C17=1.17*	*C210=0.6379/ C18=2.2221*	*C211= -0.7106/ C19= -4.448*	*C212=- 0.1998*
	in2 mf4	*C213=0.005624*	*C214=-1.29*	*C215= -1.312*	*C216=- 1.08*

From the comparison of the signals corresponding to the reality at the output of the fuzzy controller and the one to be reproduced it follows that in the periods of occurrence of free transition components (change of a specified signal at the loop inputs), the output signal of the fuzzy controller corresponds to the desired one only by sign and not by value. In the moments of changing the specified signal, the loop dynamic is mainly formed by the classic controllers of active current and speed with the depression of the circulating electromotive force (EMF). Under the transition to the forced values of controlled variables, real and desired output voltages of the PD fuzzy controller coincide. Thus, in the forced mode the fuzzy controller will significantly affect the dynamic processes in the control loops.

An analysis of transition processes shows that inclusion of a fuzzy controller parallel to a classical one results not only in an increase of a loop speed, but also in an increase of deregulation as compared with the modular optimum tuning. In addition to the increased deregulation in the time of the second matching the increased oscillation is observed. To eliminate these adverse effects, the current loop tuning coefficient increases from the value of $a_T=2$, which should be for the modular optimum to the value of $a_T=3$. This decreases the control system speed practically to the value of the modular optimum tuning. Simultaneously, the deregulation and loop speed oscillation significantly decrease to the values less than at modular optimum tuning.

For the quantitative comparison of the transition process quality we introduce the integral estimate IK, which takes into account the rate of attenuation and the value of rotation frequency deviation in the aggregate. We calculate not only the frequency deviation from the specified level of x, but also the frequency deviation to the third derivative inclusively from the frequency deviation of a motor.

$$I_K = \int_0^\infty \left(\begin{array}{l} x^2 + a_c \cdot a_T \cdot T_\mu \cdot \dot{x}^2 + \\ + a_c \cdot a_T^2 \cdot T_\mu^2 \cdot \ddot{x}^2 + a_c \cdot a_T^3 \cdot T_\mu^3 \cdot \dddot{x}^2 \end{array} \right) dt \quad (6)$$

This estimate characterizes the transition process approach to the extreme, which is determined by the solution of a differential equation of the optimized loop characteristic polynomial of the motor rotation.

$$a_c \cdot a_T^2 \cdot T_\mu^3 \cdot \dddot{x} + a_c \cdot a_T^2 \cdot T_\mu^2 \cdot \ddot{x} + $$
$$+ a_c \cdot a_T \cdot T_\mu \cdot \dot{x} + 1 = 0 \quad (7)$$

For the rigid control system, we accept the value of integral estimate I_K as the base value to which we refer all other values of this estimate.

Fig. 5 shows the calculated oscillograms of changing the motor frequency and the corresponding values of the integral estimate I_K, depending on the number of attached tubes and fuzzy controller tunings.

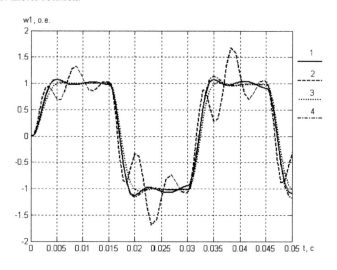

Figure 5. The curves of changing the motor shaft rotation frequency depending of the control system dynamics correction of the fuzzy controller, where: 1 – the PD fuzzy control of 0 tubes ($I_k = 1.08$), 2 – the PI fuzzy control of 0 tubes ($I_k = 1.18$), 3 – the PD fuzzy control of 4 tubes ($I_k = 1.02$), 4 – extreme ($I_k = 1$)

The values of the integral estimate I_K – the indicator of the transition process in the control jump and perturbation jump modes – were calculated for the jump of control and perturbation jump depending on the number of attached drill rods in the drilling tube and on the dynamic characteristics of a fuzzy controller.

As it follows from the analysis of obtained values, the greatest difference from the optimally tuned control system is observed when applying the PI fuzzy controller, and the least – when applying the PD fuzzy controller. The PID controller is characterized by the intermediate values of the control quality.

4 CONCLUSIONS

The conducted research allowed us to establish the following regularities:

1. In the process of modernization, replacement of a thyristor DC drive in a rotator by a transistor AC drive with a pulse-modulated impulse of autonomous voltage inverter will change the dynamic characteristics of the whole electromechanical system, since in the latter case, a bandwidth of the electric drive control system is increasing. The problem of frequency compatibility between drive systems and dynamic parameters of drilling rigs arises.

2. Increasing the depth of drilling, performed by three and more drill rods, and also the length of a rod significantly reduces the natural

101

frequencies of the polyresonance phenomena in the mechanical part of the drilling tube which with the use of an AC drive get into the bandwidth of the current and rotation frequency loops.

3. Compensation of the influence of electromechanical and electromagnetic time constants of control loops and back EMF of the motor is possible by using classic controllers.

4. The quantitative and qualitative influence of the transmission elastic properties of a drilling tube on the current and rotation frequency loops is conditioned by the fact that in the control object of the current loop there appears an uncompensated dynamic unit with four zeros and poles in the transfer function and in the speed loop – with two zeros and poles. Moreover, coefficients of the polynomials of additional transfer functions depend on the tube mass, which varies according to the number of attached rods in the drilling tube leading to the displacement of zeros and poles in the complex plane of the transfer function.

5. Compensation of the influence of additional unit dynamic properties on the dynamics of work of the current loop in the field of classical controllers leads to the need of using an additional controller with adaptive tuning and the need of determining the derivatives up to the second order, which, under the presence of noise in the signals of regulation, will adversely affect the loop control process.

6. The introduction of a fuzzy PD controller from errors at inputs of proportional-integral controllers of speed and current to the controller's output, makes it possible to preserve the advantages of active control systems with consequent correction and to effectively depress the fluctuations in the control system.

7. The quality of transients at various membership functions, does not change significantly, therefore it is advisable to choose the triangular or Gaussian membership function due to the simplicity of their implementation.

8. It is necessary to choose the output membership functions by the Sugeno type function of the zero-order as with the linearly varying functions the elastic fluctuations in the control system are poorly depressed.

9. A fuzzy controller badly reproduces a required output signal with rapid changes of the specified and perturbing signals at the beginning of each step-wise action, when classical controller compensate the inertial influent of electromagnetic and electro-mechanical time constants. When a control system works in a forced mode, classical controllers work inefficiently, depressing elastic fluctuations of a fuzzy controller. Thus, the temporary selection modes of classical and fuzzy controllers are realized.

10. With the absence of the oscillatory component of the tube's elastic vibrations in the control system signals, the fuzzy controller does not influence the dynamics of the control system.

11. Changing the natural frequencies of elastic vibrations of a drill tube does not significantly affect the quality of transients, if the tuning of a fuzzy controller is made to compensate elastic vibrations with a minimum number of rods the tube.

REFERENCES

Quang N.P. & Dittrich J.-A. 2010. *Vector Control of Three-Phase AC Machines: System Development in the practice (Power Systems).* New York: Springer: 340.

Novotny D.V., Lipo T.A. 1996. *Vector control and dynamics of AC drives.* New York: Oxford science publications: 450.

Boldea I. & Nasar S.A. 1992. *Vector Control of AC Drives (Handcover).* New York: CRC Press Taylor & Francis Group: 240.

Boldea I. & Nasar S.A. 2006. *Electric drives.* New York: CRC Press Taylor & Francis Group: 522.

Khilov V.S., Beshta A.S., Zaika V.T. 2004. *Experience of using frequency-controlled drive drill rigs in quarries in Ukraine* (in Russian). Moscow: Mountain information-analytical bulletin of Moscow State Mining University, Issue 10: 285-289.

Pivnyak G.G., Beshta A.S., Khilov V.S. 2005. *AC drive system for actuator's power control.* XIII International Symposium on Theoretical Electrical Engineering, Vol. 5: 368-370.

Khilov V.S., Plakhotnik V.V. 2004. *Estimate the natural frequencies of the drill rod under unsteady conditions* (in Russian). Dnipropetrovs'k: NGU: Proc. Scientific Transactions NGU, Issue 19: 145-150.

Pivnyak G.G., Beshta O.S., Khlov V.S. 2003. *Building Principles of Control systems for electric rotating machine cutting drilling pool* (in Ukrainian). Kharkov: NTU "KPI": Bulletin of the National Technical University "Kharkiv Polytechnic Institute", Issue. 10: 141-143.

Khilov V.S. 2006. *The changing dynamics driving the spinner rig in the application of the AC drive system* (in Russian). Krivoy Rog:

Development of ore deposits. Bulletin of KTU, Issue 1: 180-184.

Pivnyak G.G., Beshta O.S., Khlov V.S. 2005. *Drive system The round-trip operations drills* (in Ukrainian). Kharkov: NTU "KhPI": Scientific Transactions of the National Technical University "Kharkiv Polytechnic Institute", Issue 45: 223-225.

Khilov V.S. 2006. *Application of computer-aided drives in new generation boring rigs for open pit's in Ukraine.* Dnipropetrovs'k: NGU: Scientific herald NMU, Issue 5: 72-76.

Pivnyak G.G., Beshta O.S., Hilov V.S. 2004. *Adaptive fuzzy power controller to manage the drilling* (in Ukrainian). Technical electrodynamics, Issue 6: 47-52.

Khilov V.S. 2011. *Drill spinner drive dynamic performances correction of blast hole boring rig.* The materials of the international conference of "Miners Forum 2012". Dnipropetrovs'k: NGU: 90-95.

Khilov V.S. 2012. *The information-analytical characteristics of the busbar field parameters* (in Russian). The materials of the international conference of "Miners Forum 2012". Dnipropetrovs'k: NGU: 90-95.

Terekhov V.M. 2001. *Algorithms for fuzzy controllers in electrical systems* (in Russian). Electricity, Issue 12: 55-63.

Terekhov V.M., Baryshnikov A.S. 1996. *Stabilization of slow-moving traffic drives on the basis of Fuzzy-logic* (in Russian). Electricity, Issue 8: 61-64.

Zadeh L.A. 1976. *The concept of a linguistic variable and its application to the adoption of approximate solutions* (in Russian). Academic Press: 165.

Yager R. & Filev D. 1984. *Essentials of Fuzzy Modelling and Control.* USA: John Wiley & Sons: 387.

Leonenkov A.V. 2003. *Fuzzy Modeling with MATLAB environment and fuzzy TECH* (in Russian). Saint Petersburg: BHV-Petersburg, 736.

Medvedev V.S., Potemkin V.G. 2002. *Neural Network MATLAB 6* (in Russian). Moskow: DIALOG-MEPI: 496.

Processes in distribution mains during the breaks of the aerial power line.

F. Shkrabets & A. Grebenyuk
State Higher Educational Establishment "National Mining University", Dnipropetrovs'k, Ukraine

ABSTRACT: Principles of emergency currents formation and zero-phase-sequence parameters for possible types of ground fault are stated when phase conductor of aerial power line breaks in three-phase distribution networks with isolated neutral line and also with neutral line grounded through arc-suppression coil and resistor.

1 INTRODUCTION

Such kind of a failure as a wire break of one of the aerial power transmission line (PTL) phases in open-pit distribution mains makes up about 10% of all types of damages. Danger degree of this emergency state is characterized by emergency state transfer into single-phase earthing when the broken wire comes in contact with earth. Besides, depending on now fat the wire breakage point is from the PTL support, there are three modes of shortening:

- ground fault from the source (classic single-phase ground fault);
- ground fault from the electrical receiver side (only the wire connected with a consumer touches the ground);
- ground fault in two points from the sides of the source and the electrical receiver.

The first case is related to common and quite investigated types of single-phase ground faults. The second case is characterized by the fact that the mains are connection the ground through electrical receiver resistance. The third case is possible only when the breakage takes place in the middle of PTL phase wire flight.

Values of single-phase ground fault currents in distribution mains with voltage equal to 6-35 kV are determined, basically, by voltage level, mains insulation parameters (capacity and active resistance) relatively to the ground and by emergency state peculiarities. In turn, mains insulation parameters are distributive and depend on a number of factors the basic of which are electrical circuit performance (air or cable lines) and the physical parameters of lines (cross-section of conducting filaments, length, suspension height etc.).

2 THE PAPER AIMS TO:

1. To consider currents of single-phase ground faults in distribution mains with voltage of 6-35 kV during aerial PTL wire breaks for various types of neutral grounding.

2. Consider currents and zero voltages for various types of neutral grounding.

3 MATERIALS FOR RESEARCH

The most widespread and available method of emergency modes research in power-supply systems is mathematical modeling based on the processes formalization and plotting of particular mathematical models (Bernas & Zech 1982; Wilgame & Waters 1959). When constructing equivalents for distribution mains in order to research characteristics currents ground faults of during the settled mode for various types of ground faults and for the phase wire breaks, the following assumptions must be made (Pivnyak & Shkrabets 1993):

- during the normal working mode, the three-phase system is symmetric;
- longitudinal voltage drop in lines is neglected, that is the circuit phases voltage relatively to the ground is assumed identical in any circuit point;
- distributive parameters of circuit insulation relatively to ground is substituted by concentrated parametrs;
- active resistances and inductances of lines phase wires, and also inductance and active resistance of power transformer windings are equal to zero;
- interphase circuit capacities not influencing currents values and zero voltages in the settled mode are not considered.

- transient resistance in the phase ground fault point is only active.

In case of one phase ground fault from the electrical receiver when the wire breaks, the equivalent scheme with insulated neutral is shown in Fig. 1 that considers in its structure only those elements and connections that significantly influence, the emergency currents under investigation.

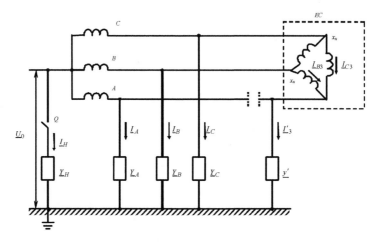

Figure. 1. Substitution of distribution circuit for ground fault occurring from electrical receiver

The scheme is compiled in accordance with the above-mentioned assumptions and restrictions, and allows to research emergency currents and characteristics of zero voltages. In order to simplify mathematical calculations, the insulation conductivity relatively to ground of the damaged line area, beyond damage point, is accepted to be equal to zero. The following markings are used in the presented scheme of substitution:

$\underline{Y}_H$ – conductivity of neutral point of circuit relatively to ground; $\underline{Y}_A, \underline{Y}_B, \underline{Y}_C$ – conductivities of A, B and C phases respectively relatively to ground; y – conductivity of transient resistance in fault point;

$\underline{U}_0$ – voltage of circuit neutral dislocation or zero voltage; $\underline{U}_A, \underline{U}_B, \underline{U}_C$ –phase voltages of feeding transformer (source); $\underline{I}_H$ – current through resistance of circuit neutral point relatively to ground; $\underline{I}_A, \underline{I}_B, \underline{I}_C$ – currents through conductivities relatively to ground of corresponding phases; $\underline{I}_3$ – single-phase ground fault current; $x_H = \omega L_H$ - inductive resistance of electrical receiver interphase winding (active resistance of electrical receiver windings is neglected considering their relatively small values); $\underline{I}_{B3}$ - ground fault current of B phase

through load resistance and transient resistance r in damage point; $\underline{I}_{C3}$ -ground fault current of C phase through load resistance and transient resistance r in breakage point.

In general case, circuit phases' $\underline{U}_A, \underline{U}_B, \underline{U}_C$ voltage relatively to the ground, circuit phases relatively to the neutral of the system $\underline{U}_{\phi A}, \underline{U}_{\phi B}, \underline{U}_{\phi C}$ and voltage of neutral dislocation $\underline{U}_0$ are equal to the value of the potential of system neutral relatively to ground and connected by the following ratios (Wilgame & Waters 1959; Sirota, Kislenko & Mikhaylov 1985):

$$\underline{U}'_A = \underline{U}_{\phi A} + \underline{U}_0; \quad \underline{U}'_B = \underline{U}_{\phi B} + \underline{U}_0; \quad \underline{U}'_C = \underline{U}_{\phi C} + \underline{U}_0$$

When aerial PTL wire breaks, the ground fault is likely to take place between the support points from both the source and the electrical receiver simultaneously. Although such damage is unlikely to occur, the circuit parameters and transient resistances influence on fault current characteristics and also on voltages and zero currents should be considered to evaluate behavior of existing protection devices during such damage. Distribution mains substitution scheme that corresponds to the examined damage is shown in Fig.2. The presented scheme shows that in such an emergency mode there are two different emergency currents in the system ($\underline{I}_{A3}, \underline{I}'_{A3}$) that flow through corresponding transient resistances (accepted only active) in contact points of broken wire with the ground (resistances $\underline{y}'_1$ и $\underline{y}'_2$).

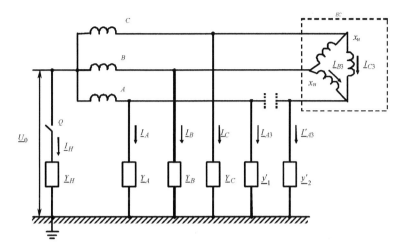

Figure. 2. Mains substitutional scheme with mixed damage for researching zero voltage

3.1 Ground fault from the source side.

If we neglect phase ground - insulation parameters (capacity and active resistance) of aerial line part after the win break point it can be assigned that ground - insulation symmetry of the entire distribution mains is not damaged. In this case when one-phase wire breaks the single-phase ground fault mode can be considered a classical type of single-phase earthling. Such type of an emergency mode within the established mode has been already considered above in detail we have also studied the influence of ground insulation parameters on the emergency current values on transient resistance in the fault point, and also the character and parameters of the ground fault mode of the distribution mains neutral.

3.2 Ground fault from the electrical receiver

In accordance with Kirhgof's first law for the insulated neutral circuit ($\underline{Y}_H = 0$), the scheme presented in Fig .1 satisfies the following:

$$\underline{I}_A + \underline{I}_B + \underline{I}_C + \underline{I}_{B3} + \underline{I}_{C3} = 0 \tag{1}$$

Substituting relevant currents for their values expressed through the corresponding voltages

and resistances and considering that in the normal mode of a symmetric system the phase voltages are presented as a symmetric three-beam star, we get

$$3\underline{U}_0 \underline{Y} + 2\underline{U}_0 \underline{y}' + (\underline{U}_{\phi B} + \underline{U}_{\phi C})\underline{y}' = 0 \tag{2}$$

where $\underline{y}'$ – phases B and C conductivity in the damage point taking into account the winding resistance of the receiver and the transient resistance in the damage point.

Considering that $\underline{U}_{\phi B} + \underline{U}_{\phi C} = -\underline{U}_{\phi A}$ and solving equation (2) relatively to $\underline{U}_0$, we will get the formula for zero-sequence voltage during the single-phase ground fault from electrical receiver in the circuit with the isolated neutral:

$$\underline{U}_0 = \underline{U}_{\phi A} \underline{y}' / (3\underline{Y} + 2\underline{y}') . \tag{3}$$

Using substituted conductivities values in the obtained equation, we get:

$$\underline{Y} = 1/R + j\omega C \quad \text{и} \quad \underline{y}' = 1/(r + j\omega L_H)$$

and after necessary conversions, we can express zero voltage through electrical circuit and electrical receiver parameters:

$$\underline{U}_0 = \underline{U}_{\phi A} R \times$$

$$\times \frac{3r + 2R - 3\omega^2 C L_H R - j3\omega(L_H + CRr)}{(3r + 2R - 3\omega^2 C L_H R)^2 + 9\omega(L_H + CRr)^2}. \qquad (4)$$

After the corresponding conversions, we get the formula for the active value and phase (relatively to the voltage vector of the damaged phase relatively to the normal mode of the circuit work) of zero voltage during the single-phase ground fault from the electrical receiver:

$$U_0 = \frac{U_\phi R}{\sqrt{(3r + 2R - 3\omega^2 C L_H R)^2 + 9\omega(L_H + CRr)^2}}; \qquad (5)$$

$$\varphi_u = arctg \frac{-3\omega(L_H + CRr)}{3r + 2R - 3\omega^2 C L_H R}. \qquad (6)$$

If, with the significant excess of insulation resistance over the capacity one, we assume that the value of the active insulation resistance is infinite, then for dead single-phase ground fault, the zero voltage will be equal:

$$\underline{U}_0 = \underline{U}_A /(2 - 3\omega^2 C L_H), \text{ or}$$

$$U_0 = U_\phi /(2 - 3\omega^2 C L_H). \qquad (7)$$

The single-phase ground fault current from the electrical receiver during the break of aerial line phase wire will be define as:
- for the circuit with insulated neutral:

$$\underline{I}_3 = -3\underline{U}_0 \underline{Y} = -3\underline{U}_{\phi A} \underline{Y} \frac{y'}{3\underline{Y} + 2y'}, \qquad (8)$$

- for the circuit with a compensating device or a resistor in neutral:

$$\underline{I}_3 = -\underline{U}_0(3\underline{Y} + \underline{Y}_H) = \underline{U}_\phi \underline{y} \frac{3\underline{Y} + \underline{Y}_H}{3\underline{Y} + \underline{Y}_H + \underline{y}}, \qquad (9)$$

where $y' = 1/(r + j\omega L_H)$ - conductivity of the mains undamaged phases relatively to the

ground in the damage point considering electrical receiver.

The Distribution circuit substitution scheme should be presented by two annexes for research of fault current and characteristics of zero sequence currents during one-of-the-phases ground fault from the electrical receiver: the first annex corresponds to the protected (controlled) line parameters; the second – to the parameters of the entire external circuit in Fig. 3.

The zero-sequence current in the controlled line, in case a damage from the electric receiver takes place in a circuit with any neutral mode, will be defined as a sum of all currents flowing in this line (load current is equal to zero):

$$\underline{I}_0 = \underline{I}_{A1} + \underline{I}_{B1} + \underline{I}_{C1} + \underline{I}_3.$$

After substitution and conversions we obtain the expression for the zero-sequence current when the controlled line with insulated neutral gets damaged:

$$\underline{I}_0 = -3\underline{U}_0(\underline{Y} - \underline{Y}_1) = -3\underline{U}_{\phi A}(\underline{Y} - \underline{Y}_1)\frac{y'}{3\underline{Y} + 2y'}, \qquad (10)$$

where $\underline{Y}_1$ - conductivity of ground - insulation of the controlled annex single phase.

Zero-sequence current in the controlled line with insulated neutral during external ground faults from the electrical receiver (the line own current) can be presented as:

$$\underline{I}_{0C} = 3\underline{U}_0 \underline{Y}_1 = 3\underline{U}_{\phi A} \underline{Y}_1 \frac{y'}{3\underline{Y} + 2y'}. \qquad (11)$$

From the expressions (9-11) we see that, like as in other cases of single-phase ground faults in mains with insulated neutral, the single-phase ground fault current and zero-sequence currents with damages from electrical receiver side are determined by zero voltage and entire circuit insulation conductivity, external circuit and controlled annex correspondingly.

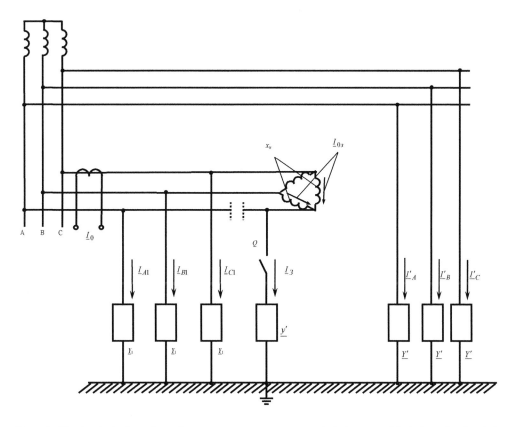

Figure 3. Circuit substitution scheme for zero-sequence current research during ground fault from the electrical receiver

If we neglect the impact of the active resistance of the mains and annex insulation, then for the dead phase grounding from the electrical receiver the ground fault current and the related zero-sequence currents are defined as:

$$I_3 = 3U_\phi \omega C \big/ \big(2 - 3\omega^2 CL_H\big), \qquad (12)$$

$$I_0 = 3U_\phi \omega \big(C - C_1\big) \big/ \big(2 - 3\omega^2 CL_H\big), \qquad (13)$$

$$I_{0C} = 3U_\phi \omega C_1 \big/ \big(2 - 3\omega^2 CL_H\big). \qquad (14)$$

Expressions (12-14) significantly differ from the known expressions for the corresponding currents during the classic single-dead fault (from the source side). Evaluation of change associated with the amplitude and phase values of emergency currents, voltages and zero-sequence currents during damages from the electrical receiver in respect to parameters of distribution mains and transient resistances in the damage point is vital for assessment of existing protection

(against asymmetric damage) devices efficiency and is necessary for the development of new performance principles and protection devices. In case an arc-suppressing reactor is fitted into the circuit neutral or high-value resistor, expressions for zero-sequence voltage, fault current and zero-sequence currents during single-phase ground fault from the electrical receiver side will be:

$$\underline{U}_0 = \underline{U}_{\phi A} \, \underline{y}' / (\, 3\underline{Y} + \underline{Y}_H + 2\underline{y}'\,) \,,$$

$$\underline{I}_3 = -\underline{U}_0 (3\underline{Y} + \underline{Y}_H) =$$

$$-\underline{U}_{\phi A} (3\underline{Y} + \underline{Y}_H) \frac{\underline{y}'}{3\underline{Y} + \underline{Y}_H + 2\underline{y}'},$$

$$\underline{I}_0 = -\underline{U}_0 (3\underline{Y} - 3\underline{Y}_1 + \underline{Y}_H) =$$

$$-\underline{U}_{\phi A} (3\underline{Y} - 3\underline{Y}_1 + \underline{Y}_H) \frac{\underline{y}'}{3\underline{Y} + \underline{Y}_H + 2\underline{y}'},$$

$$\underline{I}_{0C} = 3\underline{U}_0 \underline{Y}_1 = 3\underline{U}_{\phi A} \underline{Y}_1 \frac{\underline{y}'}{3\underline{Y} + \underline{Y}_H + 2\underline{y}'},$$

These equation do not cancel correctness of the earlier made generalizations.

3.3 Ground fault from the source and the electrical receiver during aerial PTL wire break.

Having conducted necessary conversions for substitution circuit scheme presented in Fig. 3. for simultaneous single-phase ground fault in two different points (from the source and the electrical receiver during wire breakage) we will get:

- for zero-sequence voltage in the insulated neutral circuit and with inductance or resistor in neutral

$$
\underline{U}_0 = -\underline{U}_{\phi A}(\underline{y}'_1 - \underline{y}'_2) /
$$
$$
/(3\underline{Y} + 2\underline{y}'_2 + \underline{y}'_1)
$$

$$
\underline{U}_0 = -\underline{U}_{\phi A}(\underline{y}'_1 - \underline{y}'_2) /
$$
$$
/(3\underline{Y} + \underline{Y}_H + 2\underline{y}'_2 + \underline{y}'_1)
$$
(15)

- for full current of single-phase ground fault in the circuit with insulated neutral:

$$
\underline{I}_з = \underline{I}_{Aз} + \underline{I}'_{Aз} = 3\underline{U}_0\underline{Y} =
$$
$$
= 3\underline{U}_{\phi A}\underline{Y}\frac{\underline{y}'_1 - \underline{y}'_2}{3\underline{Y} + 2\underline{y}'_2 + \underline{y}'_1} ,
$$
(16)

or

$$
\underline{I}_з = 3\underline{U}_{\phi A}\underline{Y}\frac{\underline{y}'_1}{3\underline{Y} + 2\underline{y}'_2 + \underline{y}'_1} -
$$
$$
- 3\underline{U}_{\phi A}\underline{Y}\frac{\underline{y}'_2}{3\underline{Y} + 2\underline{y}'_2 + \underline{y}'_1}
$$

and for the circuit with compensating device or with resistor in neutral

$$
\underline{I}_з = \underline{U}_{\phi A}(3\underline{Y} + \underline{Y}_H)\frac{\underline{y}'_1}{3\underline{Y} + \underline{Y}_H + 2\underline{y}'_2 + \underline{y}'_1} -
$$
(17)
$$
- \underline{U}_{\phi A}(3\underline{Y} + \underline{Y}_H)\frac{\underline{y}'_2}{3\underline{Y} + \underline{Y}_H + 2\underline{y}'_2 + \underline{y}'_1}
$$

- for zero-sequence current correspondingly for circuit with insulated neutral and with inductance or resistor in neutral

$$
\underline{I}_0 = -3\underline{U}_0(\underline{Y} - \underline{Y}_1) =
$$
$$
= 3\underline{U}_{\phi A}(\underline{Y} - \underline{Y}_1)\frac{\underline{y}'_1 - \underline{y}'_2}{3\underline{Y} + 2\underline{y}'_2 + \underline{y}'_1} ,
$$

$$
\underline{I}_0 = -\underline{U}_0(3\underline{Y} - 3\underline{Y}_1 + \underline{Y}_H) =
$$
(18)
$$
= \underline{U}_{\phi A}(3\underline{Y} - 3\underline{Y}_1 + \underline{Y}_H)\frac{\underline{y}'_1 - \underline{y}'_2}{3\underline{Y} + \underline{Y}_H + 2\underline{y}'_2 + \underline{y}'_1} ;
$$

- for the controlled annex's own current we get:

$$
\underline{I}_{0C} = 3\underline{U}_0\underline{Y}_1 =
$$
$$
= -3\underline{U}_{\phi A}\underline{Y}_1\frac{\underline{y}'_1 - \underline{y}'_2}{3\underline{Y} + 2\underline{y}'_2 + \underline{y}'_1}
$$

$$
\underline{I}_{0C} = 3\underline{U}_0\underline{Y}_1 =
$$
(19)
$$
= -3\underline{U}_{\phi A}\underline{Y}_1\frac{\underline{y}'_1 - \underline{y}'_2}{3\underline{Y} + \underline{Y}_H + 2\underline{y}'_2 + \underline{y}'_1} ,
$$

where $\underline{y}'_1 = 1/r_1$ - conductivity in the damage point from the power source;

$\underline{y}'_2 = 1/(r_2 + j\omega L_H)$ - conductivity from the electrical receiver in respect to B and C phases, considering the load.

Amplitude and frequency characteristics of the ground fault emergency currents are of practical interest in the transition mode. The authors' studies and publications analysis show that transition processes are characterized by occurrence of (except free mid-frequency vibrations) high-frequency (dozens and hundreds of kHz) vibrations of discharge current substantiated by capacity charge change relatively to ground of damage phase. Multiplicity of transition current during ground faults in insulated neutral mains depends on the fault moment, on active phase voltage value, to some extent on the circuit ground parameters and on transient resistance value in the fault point. With increase of the resistance, the amplitude of transient current rapidly decreases.

Parameters of the circuit groundinsulation and transient resistance in the damage also point influence transition process duration. If the transient resistance in the ground fault point is 100÷200 O, the transition process in the circuit with insulated neutral transfers from periodic damping into aperiodic (Srota 1974). For metallic ground faults in insulated neutral mains, we can consider that for the real parameters of the circuit the transition process practically stops in 10...15 ms.

In mains with compensated neutral depending on the distribution mains parameters ratio (including parameters of compensating device), the transition process can have vibrational or aperiodic character.

Maximum of transient current free component in the circuit with compensated neutral as well as in the circuit with insulated neutral can exceed the amplitude of the settled current by the number of times defined by the ratio of the circuit full ground insulation resistance to transient resistance in the fault point.

Transition process in the circuit with transistor in the neutral practically does not differ from transition process in the circuit with insulated neutral. However, switching resistor into the circuit neutral leads to a rapid decrease of active resistance of the circuit ground insulation and accordingly to increase of transition process damping coefficient due to which transition process duration is reduced.

To exclude occurrence of emergency modes in open-pit distribution mains related to the aerial PTL phase wire breakage and prevent further failure, it is proposed to use the principle of protective (preventive) shutdown.

Mentioned measure is capable of preventing the ground fault mode of a single-phase wire of aerial PTL due to disconnecting the line from power source until broken wire contact with the ground. For evaluating the realization of safety (preventive) shutdown under conditions of open-pit distribution mains it is necessary to research time intervals connected with breakage processes detection and shutdown of working voltage supply to emergency line from one side, and process of a broken wire movement until the moment of its contact with the ground or an equipment. In general, emergency line shutdown time in case of phase wire breakage is defined as follows:

$$t_{om\kappa} = t_{c.3} + t_{o.3} + \Delta t ,\qquad (20)$$

where $t_{c.3}$- intrinsic time of protective (preventive) device shutdown that is spent on the breakage identification and formation of control command (up to 0.1 ms); $t_{o.6}$ – response time of power switch defined at the beginning of the line at open-pit substation or at open-pit distribution point and makes up not more than 0.2 s; Δt – possible time delay of protective (preventive) shutdown. The mentioned time is accepted to be zero taking into the account that open-pit lines

are related to, as a rule, the last step of energy distribution.

The time interval from phase wire breakage until its contact with the ground (time of the wire dropping) is defined as:

$$t = \left(2 \cdot h \cdot g^{-1}\right)^{0,5} ,\qquad (21)$$

where h – phase wire support height of an open-pit distribution circuit with voltage of 6 kV (6 m); g –free fall acceleration.

The implemented calculations for the given physical parameters aerial PTL have shown that the time interval from the wire breakage moment till its contact with the ground is 1.1 s, and the time of an emergency line shutdown will make up not longer than 0.3 s. It follows that there is a real possibility to prevent ground fault mode with the help of special protective device from aerial PTL wire breakage.

One of the proposed methods to detect aerial PTL wire breakage is de-energizing of the damaged phase. It is necessary to analyze the following possible reasons of short-term absence of working currents in one of the phases in the line:

- wire breakage of one PTL phase;
- simultaneous wires breakage of PTL two phases;
- asymmetrical shutdown of the line by power switch (various time index of contacts disconnection);
- asymmetrical switching of a load line (various time index of contacts fault);
- open phase-mode of distribution circuit.

4 CONCLUSIONS

1. With damages from the electrical receiver in mains with insulated neutral the active value of zero-sequence voltage and emergency currents for real parameters of the circuit insulation and winding resistance of electrical receivers is two times lower of the corresponding voltage at faults in the same circuit from the power source.

2. Character and degree of influence of the circuit ground insulation parameters and values of transient resistance in the damage point on zero-sequence voltage value and currents is analogical to the given parameters influence during classic type of single-phase ground fault.

3. Phase characteristics of zero-sequence parameters for the classic type of faults at other things being equal differ by dislocation of corresponding vectors for near 180°.

4. At phase ground fault from the source and electrical receiver simultaneously, active values and positions of fault current vectors, and also currents and zero-sequence voltage are defined basically by ratio of transient resistances in damage points. Phase characteristics of voltage vectors zero-sequence current at mixed damage can theoretically change within the range of 270° in mains with insulated neutral and within the range of 360° in mains with compensated neutral.

REFERENCES

Bernas, S. & Zech, Z. 1982. *Mathematical models of electroenergy systems elements* (in Russian). Moscow: Energy issue: 312.

Wilgame, R. & Waters, M. 1959. *Neutral grounding in high-voltage systems* (in Russian). Moscow: State energy issue: 415.

Pivnyak, G.G. & Shkrabets, F.P. 1993. *Assymetrical damages in electrical mains of open-pits: reference book* (in Russian). Moscow: Nedra: 192.

Samoylovich, I. S. 1976. *Neutral modes in electrical mains of open-pits* (in Russian). Moscow: Nedra: 175.

Sirota, I.M., Kislenko, S.N. & Mikhaylov, A.M. 1985. *Neutral modes of electrical mains* (in Russian). Kyiv: Scientific thought: 264.

Sirota I.M. 1974. *Influence of neutral in mains of 6-35 kV on safety conditions* (in Russian). Neutral modes in electrical systems: Kyiv: Scientific thought: 84-104.

Energy Efficiency Improvement of Geotechnical Systems – Pivnyak, Beshta & Alekseyev (eds)
© 2013 Taylor & Francis Group, London, ISBN 978-1-138-00126-8

Operating dynamics of parameters and technical losses in the components of power supply systems

P. Krasovskiy, D. Tsyplenkov & O. Nesterova

State Higher Educational Institution "National Mining University", Dnipropetrovs'k, Ukraine

ABSTRACT: This study formulates the main reasons for changes in time under the long term idle capacity operation as well as presents the research results on the effects of aluminum corrosion on electric parameters of stationary transmission lines and energy losses.
Key words: energy losses, idle capacity, power transformer, aluminum corrosion.

1 INTRODUCTION

In connection with the sharp decrease of investments in the development of power grids and technical re-equipment, development of mode control systems and electric power assessment, there emerged the tendencies, having a negative influence on the level of grid losses. These tendencies are as follows: out-of-date equipment, physical and moral depreciation of power supply systems components, non-compliance of installed equipment with the transmitted load. From the above said it follows that against the background of ongoing changes of the economic mechanism in the energy sector and economic relations changes in the country, the problem of reducing power losses in power grids hasn't lost its topicality, but on the contrary, has distinguished itself as one of the tasks ensuring the financial stability of energy supplying organizations.

Electric power transmission and distribution losses are accepted as a standard basis for determining the economically substantiated technological power consumption (Vorotnickiy & Kalinkina 2003). In urban power distribution networks, the idling loss (XX) of transformers Pxx (Kazakov, Kozlov & Korotkov 2006) is a major component of energy losses (up to 30%). A decrease of transformer load factors, due to electric power redistribution, increases the share of the XX losses in the total transformer losses. When calculating energy balance, the transformer losses P_{xx} are accepted as equivalent to the certified value $P_{xx\ pasp}$. In practice, the certified value $P_{xx\ pasp}$ does not always correspond to the actual transformer losses and for different transformers the variation can be considerable.

The inexact specification of P_{xx} results in a substantial error in the calculations of electric power distribution. It is possible to assert that the energy losses in power transformers change in time and the dynamics of these changes depends at least on the operation life and conditions and also on the kinds and amounts of transformer failures and the quality of repair.

The long-term exploitation of transmission lines causes increased energy losses even if the load remains steady. One of the reasons is the configuration change of separate power lines (PL) sections and, as a result, equivalent resistance values. The main possible reasons for changes of the PL longitudinal resistances are as follows:

– decreasing the cross-sectional area and increasing the length of wires, caused by their permanent deformation as a result of wind, glaze and other types of loads;

– corrosion of the PL wires caused by different climatic factors including acid rains, humidity, high temperature and sun radiation also leads to the decrease of their active cross-section;

– increasing the specific resistance of the material of the PL wires, caused by the change of their structure (aging) and residual deformation („peening");

– deterioration of the technical condition of insulators causes relatively high leakage currents.

The purpose of the paper is:

– to determine the structure of the XX losses in transformers;

– to consider the current understanding of physical processes flowing in electrical steel under the operating conditions;

– to analyse the factors influencing the structural constituents of losses;

– to expound the research results on 'the impact of aluminum corrosion on electric parameters of stationary overhead transmission lines'.

2 THE MAIN PART

2.1 The research results on the dynamics of technical losses in power transformers

A failure analysis of power transformers showed that most of the failures are typical for all types of transformers regardless of the operating conditions (up to 80% of the total failures). Operation features effect only the possibility of an earlier rise of a failure and the severity of failures (Kazadzhan 2000). We will consider the main failures and operating modes, leading to the increased idling losses P_{xx} due to the deterioration of the steel core magnetic properties (steel aging processes) and reduction of the transformer oil quality.

2.1.1 Damages of a magnetic circuit

1. Imperfectness of intersheet insulation arising from overheating, produced by eddy- or short-circuited currents, formed as a result of active steel insulation failure in contact areas with coupling pins, presence of nicks etc., and also grounding violations. Moisture which is condensed on the oil surface, gets to the upper yoke, penetrates between the active steel plates in the form of oil-water sludge (mixture of moisture and hot oil), destroys the intersheet insulation and causes steel corrosion.

2. Local shorting of steel plates and "conflagration" in steel because of the presence of some foreign metallic or conductive particles closing steel plates in this place, insulation failures of coupling pins creating a short circuit squirrel cage of touching two parts (any metal part and a rod) in two points of contact.

3. Insulation failure of steel plates causing a shorting of steel plates, wrong grounding creating a short circuit squirrel of destruction or absence of insulating spacers in the joints of the joint magnetic circuit.

4. Insulation failure of the magnetic circuit steel plates due to the heightened magnetic circuit vibration caused by weakening of pressing the magnetic circuit, spontaneous unbolting and free motion of fixing details, motion of coming off margin steel plates in rods

or yokes, weakening of pressing joints, disruption or destruction of insulating spacers in joints.

2.1.2 Damages of windings

1. Winding short circuits arising from the destruction of winding insulation caused by the natural wear-and-tear or continuous overloads with insufficient cooling, from the insulation deterioration of coils due to mechanical damages resulting from the pushes or winding deformations at short circuits and other emergency modes as well as from the outcrop of windings due to the low oil level, wire insulation defects missed when making windings.

2. Breaks in the windings because of burning-out of outputs as a consequence of electrodynamic forces at short circuits or loose connections and of part of coils as a result of short circuits in the winding.

3. Breakdown of the transformer body caused by the basic insulation defectiveness resulting from the aging or presence of cracks, a creeping electric discharge in the fulled-board insulation, a decrease of oil level, ingress of moisture or dirt, overvoltage, deformation of windings under short circuits.

4. Breakage of one or several parallel wires in the winding coil due to the burnout of outputs as a consequence of electrodynamic forces at short circuits or loose connections.

2.1.3 Deterioration of transformer oil properties

1. An abnormal increase of oil temperature and local heatings because of the cooling system failures, transformer overloads, internal transformer failures.

2. Deterioration of oil quality as a consequence of internal failures, accompanied by a cracking process (when gaseous products of oil decomposition are dissolved in the remaining oil) and strong release of combustible gases.

According to (Stepanov & Andreev 2011), very often, it is not taken into account that capital repair can cause increased idling losses (XX) as compared to their certified values. In some cases the increase of losses XX can be very substantial. The main reasons of this are as follows:

– mechanical effects of the magnetic circuit (strikes on steel, plates bends, throwing plates on each other, cutting of plates and rolling-up of wire-edges, pressing a magnetic circuit) on the electrical steel;

– application of electrical steels with worse magnetic characteristic as compared to the one built into a magnetic circuit, when replacing burnt-out plates;

– repairing without replacing damaged plates, when they are evenly distributed on the magnetic circuit cross-section;

– use of the old magnetic system (with the increased transformer operation period the idling losses are growing (Kaganovich & Raykhlin 1980)).

During the long-term operation of power transformers the real values of their idling power change in time for the following reasons:

1. When the insulation of steel plates is poor, there emerges a path for the closure of eddy currents between the sheets of steel in the rod, and it may cause the increase of losses XX by 10-30% (Kaganovich & Raykhlin 1980).

2. When a transformer operates for a long time, the compression of laminated core sheets by coupling pins or bandages weakens and this causes partial pressing of the transformer core, which leads to the occurrence of spurious gaps on the way of shorting the flux of mutual inductance. Because of this, the transformer current XX increases by 10% (Alekseenko, Ashryatov, Veremey & Frid 1978). At the same time, the leakage flux grows, causing the increased power losses on the ways of shorting these fluxes (in steel of a tank and other steel structural elements – by 20%). The increased XX losses, corresponding to these processes, can reach 5%.

3. Operation of a transformer involves overheating (from short-circuits, deterioration of heatsink conditions due to the aging transformer oil etc) above admissible values leads to the deterioration of the core steel magnetic properties (the long-term temperature rise contributes to the structural change called the aging process) and to the increase of current XX and the value of P_{xx}, which with the passage of time can reach up to 4% (Alekseev 2002; Molotilov, Mironov & Petrenko A.G. 1989).

4. The growth of a transformer life cycle leads to the deterioration of insulation dielectric properties of windings, outputs, transformer oil, as a result of which the insulation resistance drops, leakage currents increase and, consequently, the dielectric losses rise (they can reach 10% of the total value of idling losses in high-voltage transformers (Alekseenko, Ashryatov, Veremey & Frid 1978)).

5. Low-quality magnetic circuit reassembly during the transformer maintenance leads to the increase of the XX losses by 20%. Replacement of the steel of a core with the other brand of steel (hot-rolled with cold-rolled), direct butts with scarf joints leads to the change of losses in the magnetic circuit (Alekseenko, Ashryatov, Veremey & Frid 1978). The additional mechanical processing of electrical steel plates increases magnetic circuit losses by 5-10% (Molotilov, Mironov & Petrenko A.G. 1989). Annealing of steel plates reduces specific losses in them by 15% (Alekseenko, Ashryatov, Veremey & Frid 1978), but simultaneously worsens magnetic properties of the magnetic circuit electrical steel and increases the current XX by 20% with the growth of electric losses in primary windings.

The change of losses XX is substantially affected by the change (during the repair) of magnet data, insulating gaps, replacement of transformer oil, solid insulation of windings and outputs. If the ratio of voltage and the number of windings in the primary coil changes, the magnet flux in a transformer does the same and the transformer losses XX change proportionately to the square of the magnet flux changes. In addition, if the transformer works in the conditions of unbalanced load, the zero sting magnetic fluxes arise and, as a consequence, additional losses XX increase too.

Change of P_{xx} during the time of work is confirmed by special measurements results of 13 transformers TM 250/10 with the service period from 2 to 34 years (Akimov 1946). The transformer with the longest service life was characterised by the maximal losses XX. The data showed that real but not passport loss values of transformers XX, corresponding to the terms of their service, should be taken into consideration when calculating the standard of losses for distribution networks with upper-voltage of 6-10 kilovolt.

In (Kazakov, Kozlov & Korotkov 2006) the test data analysis of 143 different 100-400 kilovolt-ampere transformers was conducted, the secondary winding voltage was 6-10 kilovolt and the service period was from 1 to 51 years. There were transformers with P_{xx} 60% higher than $P_{xx.pasp}$. The calculations showed that real losses of all the 143 transformers exceeded the design value by 15,5%, attained from passport loss values (for transformers with the service period more than 20 years this excess made up 19%). Average growth of P_{xx} of the transformers with the service period of more than 20 years is 1,75%/y, less than 20 years – the idling losses are accepted as equivalent to the passport values with the error of up to 8%.

2.2 The research results on the dynamics of technical losses in transmission lines

Over time, the overhead transmission lines are exposed to atmospheric corrosion. In its turn, the atmospheric corrosion of metals, including aluminum, is the electrochemical corrosion. The electrochemical processes take place inside the moisture film which is, under the atmosphere corrosion, on the metal surface. As the film is not thick, the oxygen delivery to the metal surface is not complicated. It can affect the kinetics of electrode processes.

If the air, not saturated with water steam, i.e. with relative humidity below 100%, is subjected to cooling, at a certain temperature the saturation limit will be reached and the moisture will begin to release in the form of a mist. If a smooth metallic surface is in the atmosphere with relative humidity below 100%, only a unimolecular adsorption moisture film will appear on it (Gerasimov 1967).

When the temperature of water steams goes down or its amount increases, the dew point is reached in the air and the water fog is deposited on the metal in the form of small droplets. Depending on the metal, surface state and atmosphere saturation degree, the drops may be bigger or smaller. Under further depositing of vapor the drops can merge together to form a continuous thin moisture film. If the metal surface is rough or it contains solid particles of dust, coal etc and also loose areas of the protective film and products of corrosion, the condensation of moisture with the formation of a water layer takes place long before the dew point is reached in cavities, pores and cracks.

During the process of fog formation, the gases of air (N_2, O_2, CO_2) and also present in the air gas admixtures, such as SO_2, oxides of nitrogen, HCl and others will be dissolved in the fog drops, and, thus, drops and a moisture film on the metal surface won't consist of clean water but of solutions of corresponding gases in water. Being dissolved in water, a great amount of SO_2, nitrogen oxides and HCl will create noticeable concentrations in the moisture film, even if their concentration in the air is relatively small.

The kinetics of the cathode process on aluminum in the volume of 0,1-н. of the sodium chloride solution is hardly different from the one

taking place in a 165 μm-thick moisture film. When using such a film, the sharp voltage drop takes place and the cathode current density goes up (Pivnyak, Kigel' & Volotkovs'ka 2002). This is explained by the fact that the electrochemical behavior of aluminum is substantially dependent on the presence of an oxydic protective film on its surface. In case of the cathode polarization the alkalescence of medium takes place. It results in the destruction of protective oxidefilm on the aluminum and makes reverse voltage changes. As the alkalescence effect is more noticeable in the moisture film, in this case there is more sharp voltage drop.

The atmospheric corrosion speed goes up if the air humidity increases. The relative humidity value, which causes a sharp increase of the corrosion speed, is called as critical humidity.

The increase of the aluminum corrosion speed in the atmosphere polluted by a sulfurous gas is the peculiarity of the latter to be reduced on aluminum if the potential is sufficiently positive. The speed of the cathode process increases substantially (Pivnyak, Kigel' & Volotkovs'ka 2002).

The increase of the aluminum corrosion speed at the presence of chlorine is explained by the growth of the cathode process speed and increase of stationary potential. Chlorine and hypochloric acid, formed at interaction between chlorine and moisture serve as depolarizators (Pivnyak, Kigel' & Volotkovs'ka 2002).

In the non-aggressive atmosphere, the aluminum alloys are sufficiently stable; in the industrial and marine atmosphere, the alloys of aluminum are mainly exposed to the pitting corrosion.

The aggressiveness of the marine atmosphere depends on the direction of predominating winds, precipitation amount, air humidity, amount of salt water sparks reaching the metal surface.

Hydrate of aluminum oxide $Al(OH)_3$ formed at the corrosion, may further change with the formation of aluminum oxide (Gerasimov 1967):

$$4Al(OH)_3 > 2Al2O3 + 6H2O. \qquad (1)$$

It is important for power services to estimate the influence of aluminum corrosion of stationary overhead transmission lines on the grid electric parameters, in particular, on the line losses.

The corrosive process causes the diameter decrease of the conductor d. We will ignore the unevenness of corrosive processes along the full length of the power line. Then it is possible to say that aluminum corrosion of wires in power lines causes the undergage of aluminum thread on the value Δd_0 per year. In its turn, the wire diameter value depends on the power line resistance.

As the diameter change is insignificant, it is possible to ignore the reactance change of the transmission line.

Thus, it is possible to draw up a conclusion, that the corrosion of aluminum wires of transmission lines influences the value of active resistance R, and the size of active losses P, as $P \approx R$.

In turn, the active resistance in directly proportional to the active longitudinal resistance r_0, which can be calculated using the equation:

$$r_0 = \frac{1000}{\gamma \cdot F}, \tag{2}$$

where γ – conductivity of material, for aluminum γ = 32..34 m/(Ohm·mm^2); F – wire cross-section area, mm^2 (Kazadzhan 2000).

Aluminum wire of any brand for overhead transmission lines generally consists of n aluminum threads with the diameter d_0 each (Kazadzhan 2000). Then

$$F = n \cdot F_0, \tag{3}$$

where $F_0 = \dfrac{\pi \cdot d_0^2}{4}$ – cross-section area of aluminium thread, мм^2.

The next step is to show the dependence of cross-section area of an aluminum thread on the corrosion process duration (operational period of overhead transmission lines)

$$F_{0\kappa} = \frac{\pi \cdot (d_0 - \Delta d_0 \cdot T_\kappa)^2}{4}, \tag{4}$$

where Δd_0 – undergage of an aluminium thread, mm/year; T_κ – corrosion process duration, years.

Then the dependence of active longitudinal resistance on the corrosion process duration will look like this:

$$r_{0\kappa} = \frac{1000}{\gamma \cdot F_{0\kappa} \cdot n}. \tag{5}$$

We will determine the coefficient of corrosion in such a way:

$$k_\kappa = \frac{r_{0\kappa}}{r_0}. \tag{6}$$

By substituting the expressions (5) and (2) into

(6) and performing mathematical transformations we will get

$$k_\kappa = \frac{1}{(1 - \dfrac{\Delta d_0}{d_0} \cdot T_\kappa)^2}. \tag{7}$$

Then the active-power losses in the line will be determined as

$$P = I^2 \cdot R \cdot k_\kappa = I^2 \cdot R \cdot \frac{1}{(1 - \dfrac{\Delta d_0}{d_0} \cdot T_\kappa)^2}. \tag{8}$$

The average corrosion rate of aluminum under the wet film for the period of 100 hours is:
– industrial atmosphere – 0,04 μm;
– marine atmosphere – 0,11 μm;
– rural atmosphere – 0.02 μm.

That is, if we assume that duration of humidification of the aluminum surface is 100% or 8760 hours per year, then the maximum possible speed of corrosion for different areas will be:
– industrial atmosphere – to 3.5 μm;
– marine atmosphere – 9.64 μm;
– rural atmosphere – 1.75 μm.

However, given the fact that lately the concentration of pollutants in the atmosphere affecting the corrosion rate has increased significantly, the corrosion rate of aluminum in the range of 1,75 -3,5 μm per year was adopted for research.

Figures 1 and 2 show the calculated dependences of the corrosion coefficient from the time of wires operation at extreme values of the corrosion rate Δd_{min} = 1,75 μm/year and Δd_{max} = 3,5 μm /year for:
– wires A-16 with the diameter d = 1.7 mm (Fig. 1-a and 1-b);
– wire A-95 with the diameter d = 4.1 mm (Fig. 1-a and 1-d);
– wire And-35 with the diameter d = 2.5 mm (Fig. 2-a);
– wire And-50 with the diameter d = 3.0 mm (Fig. 2-b).

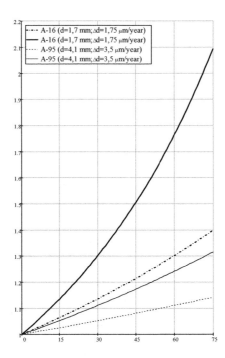

Figure 1. Dependence $k_k = f(T_k)$ for wires A-16 (wire diameter d = 1.7 mm) and A-95 (wire diameter d = 4.1 mm) at extreme values of the corrosion rate $\Delta d_{min} = 1{,}75$ **μm**/year and $\Delta d_{max} = 3{,}5$ **μm**/year.

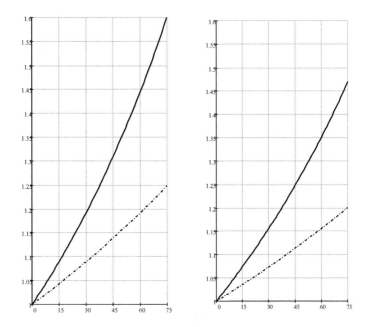

Figure 2. Dependence $k_k = f(T_k)$): – left – wires A-35; – right – wires A-50
—— wire diameter $d_0 = 2{,}5$ mm;   – · – wire diameter $d_0 = 3$ mm.

118

3 CONCLUSIONS

1. In line with the growth of the transformer service period one can observe imperfectness of magnetic circuit intersheet insulation, active steel insulation failure in contact areas with coupling pins, presence of nicks, worsening of insulation dielectric properties of windings, outputs, transformer oil etc. As a result, it leads to the increased values of eddy- and leakage currents, decreased values of insulation resistance, and thus the increased transformer losses.

2. When calculating energy balance, idling transformer losses are accepted as equivalent to the certified value, but in practice this does not always correspond to the real transformer losses and for different transformers the difference can be considerable and it is determined by the operation period and operation conditions.

3. In order to improve the quality of calculating the standard of losses for distribution networks with the upper-voltage of 6-10 kilovolt it is necessary to take into account actual but not certified values of idling losses of transformers which correspond to their operation period.

3. The long-term exploitation of overhead transmission lines is connected with the decreased wire diameter of air transmission lines and, correspondingly, increased specific wire resistance.

4. Changes of longitudinal resistance in the air transmission lines is the reason of the increased thermal losses in stationary overhead networks when their operation period increases.

REFRENCES

Vorotnickiy, V.E. & Kalinkina, M.A. 2003. *Calculation, setting of norms and decline of losses of electric power in electric networks.* Manual, 3th stereotype (in Russian). Moscow: 64.

Kazakov, Yu.B., Kozlov, A.B. & Korotkov, V.V. 2006. *Account of change of losses of idling of transformers in the period of term of service at the calculation of losses in distributive networks* (in Russian). Moscow: Electrical Engineer, Joint-stock COMPANY "Sign", Issue 5: 11 - 16.

Kazadzhan, L.B. 2000. *Magnetic properties of electrical engineerings steel and alloys* (in Russian). Moscow: LTD «Science and technologies»: 224.

Stepanov, V.M. & Andreev, K.A. 2011. *The Technical decisions on diagnostics of power transformers* (in Russian). Tula: TULGU: Engineerings sciences, Issue 6, Vol. 1: 74 – 81.

Balabin, A.A. 2009. *Increase of authenticity of calculation of losses of electric power in the transformers of 10(6) /0,4 кV* (in Russian). Mechanization and electrification of agriculture, Issue 4: 22-23.

Kaganovich, E.A. & Raykhlin, I.M. 1980. *Test of transformers power to 6300 кVA and tension to 35 кV* (in Russian). Moscow: Energy: 312.

Alekseenko, G.V., Ashryatov, A.K., Veremey, E.A. & Frid, E.S. 1978. *Test of powerful transformers and reactors* (in Russian). Moscow: Energy: 520.

Alekseev, B. A. 2002. *Kontrol' the states of large power transformers* (in Russian). Moscow: NC ENAS: 216.

Molotilov, B.V., Mironov, L.V. & Petrenko, A.G. and other. 1989. *The Kholodnokatanye electrical engineerings became: Reference book* (in Russian). Moscow: Metallurgy: 168.

Akimov, G.V. 1946. *Bases of studies about corrosion and defence of metals* (in Russian). Moscow: the State scientific and technical publishing house of literature on black and coloured metallurgy: 463.

Gerasimov, V.V. 1967. *Corrosion of aluminium and his alloys* (in Russian). Moscow: Metallurgy: 114.

Pivnyak, G.G., Kigel', G.A. & Volotkovs'ka, N.S. 2002. *Calculations of electric networks of the systems of power supply* (in Ukrainian). Dnipropetrovs'k: NMU: 219.

Energy saving in electrified transport by capacity storages

A. Kolb
State Higher Educational Institution "National Mining University", Dnipropetrovs'k, Ukraine

ABSTRACT: The system of the electrified transport braking energy accumulation by capacity stores and simultaneous network unloading from the total power inactive constituents and loading voltage stabilizing is offered.

1 INTRODUCTION

The problem of the power saving and improvement of electric energy quality in the grids of alternating voltage and the traction transforming substations with the one-way energy converter, is the most topical in electrical power engineering.

Electrified transport (industrial locomotives, escalators, trams, trolley-buses, etc) is characterized by irregular schedule of electric energy consumption, which is especially intense during starting up stage and less considerable during steady motion, reaching its minimum while braking. Hence, traction substations (TS) are constantly affected by significant load fluctuations. Electric potential on TS buses goes down at peak loads which appear due to the overlapping of the striking currents of several traction means. It slows down the starting process of electric transport and leads to electric power loss, which makes it necessary to increase the total capacity of TS power facilities in order to compensate peak loads.

At the present time the kinetic energy of moving masses accumulated during traction is not recovered back into the direct current circuit during braking, but is emitted in the form of heat in brake resistors. Besides, it is also topical to increase the quality of energy for hybrid (dc and ac) system of energy delivery (for instance, joint networks of public utility services and municipal electrified transport).

2 FORMULATION THE PROBLEM

The research suggests a system to accumulate the braking energy of electrified transport by means of capacitance storage devices which ensure quick exact removal of inactive components of the gross power from the network and voltage regulation with help of pulse-width modulation (PWM) that can be controlled by the instantaneous values of jet capacity. Owing to that electric energy quality rises and electromagnetic compatibility of various customers with dynamic reactive, nonlinear and nonsymmetrical load also increases.

3 MATERIALS FOR RESEARCH

During braking process traction motors are switched into generator mode. In this case, kinetic energy changes into electric one which with the help of brake resistors is transformed into heat and is dispersed in environment. It is considered that loses can account for 30% of the whole consumed electric energy. Consequently, the economy of energy on electric transport may be achieved by utilizing the energy of braking.

The problem can be solved both by using recuperation mode and of the energy of braking. The main difficulty in using a recuperative mode of braking lies not in the fact that kinetic energy turns into electric one and returns to the overhead wiring, but in the net's ability to accept this complementary energy. It is possible when the energy is consumed from the net by other transport means in traction conditions or by its transmission into three-phase system of the alternative current. For this purpose it is necessary to transform the steady voltage into three-phase alternative which is synchronized with the net in frequency and amplitude.

All the above-mentioned drawbacks may be removed if tracking substations are supplied with capacitance energy storage devices (Brodsky 2008) which will take extra energy of breaking and subsequently return it to the overhead wiring during starting and speeding-up of trains (fig. 1).

For managing the energy quality it is expedient to use the voltage transformers with the double-sided conduction which are operated according to the instantaneous magnitudes of inactive components of the gross power (fig. 2).

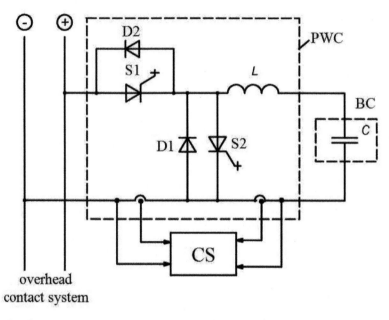

Figure 1. The functional scheme of recuperation energy accumulation at the traction substations

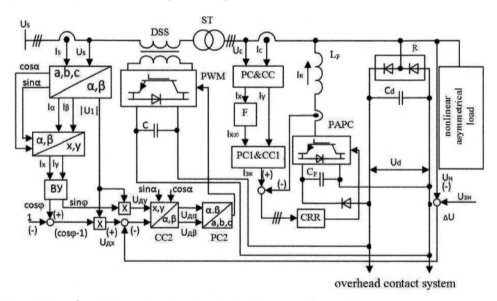

Figure 2. The functional scheme of traction substation electric energy quality control

The main elements of the energy accumulation system of regenerative braking with capacity storage devices (Fig. 1) are the battery of charge-store capacitor (BC), current-reversible pulse-width converter (PWC) with various transmission coefficients, control system (CS) with current and voltage sensors at the input and output of the transformer.

The scheme (Fig. 1) shows controlled charge of capacitor battery which considerably reduces losses of energy. The connection of capacitor battery to the overhead wiring and its recharging is exercised automatically when the voltage level on its bus bars is somewhat higher than the top rectified voltage. The controlled charge of the capacitor battery is realized by the method of pulse-duration modulation, and by means of switch S1 commutation (Fig. 1). When S1 is turned on, positive voltage impulse is formed at the output. When S1 is turned off, the energy, which is stored in the induction of the filter, is transferred to the accumulator, through D1, which is turned on to the regime of zero rectifier.

Recovery of energy, which is stored in the capacitance storage, to the overhead system occurs when its bus voltage is lower than the baseline value. Power transmission from capacitive storage to the overhead system is exercised by means of switch S2 commutation. The recovery of energy to the system is realized with the conversion coefficient bigger than one. It is achieved by dividing in time the processes of energy storage in the inductor, when S2 is closed, and its transmission to the contact system, through diode D2, when S2 is opened. Thus, the total voltage of the capacitor battery and EMF of filter self-induction is applied to the buses of the overhead system. This makes it possible to transfer energy from the capacitor battery, with the voltage much lower than in the overhead system (capacitive battery runs down to the certain voltage level). Thereby the energy efficiency of capacitive storages is considerably increased.

Three-phase linear regulators (Brooms 1985; Zhezhelenko 2000), which contain boosting transformer with a mechanical switch device, are used nowadays for voltage regulation across the high side of the power transformer of the traction substation. Filter-compensating devices (FCD), which are tuned to the frequency corresponding to the higher harmonic of the current, are used for the reactive-power compensation and harmonic filtering. Such FCD are bulky and insufficiently effective, as divergence of their real parameters from the design ones breaks conditions of current resonance. For the complex harmonic spectrum of current using of FCD is ineffective (Zhezhelenko 2000). Transformer-valve compensators of the net reactive energy, with voltage load stabilization of mini-substations (Inkov 2003), are free from certain disadvantages, however it doesn't allow balancing the load currents and it doesn't ensure qualitative harmonicity of their shape, which brings about incidental losses in transformers and the grid (Inkov 2007).

The technical solution, which combines devices capability (Kolb 2006; Kolb 2005) for

achieving reactive-power compensation of the circuit, filtration of higher harmonics of the current, load balancing with simultaneous voltage stabilization of transformer substation, is offered in this work. The solution for this problem is realized by the throttling in the wide range of amplitude and voltage phase of boosting transformer with the help of PWM. The second voltage inverter, which is a part of parallel active power compensator, is used for the compensation of inactive components of the gross power (power of shift, distortion and asymmetry). Controlling parameter of both voltage inverters (Fig. 2) is formed with the use of instantaneous values of inactive current components to be compensated and controlled on the high and low sides of the supply transformer (ST) of the traction substation.

On the basic of the above questions is realized management in the closed loop automatic system of the gain-phase voltage regulation DSS for reactive power compensation in the network and stabilization of the traction substation voltage. In the scheme fig. 2 PDM is connected straight on dc voltage of the traction network. Smooth control of consumed or generated energy implements by the lagging and advanced phase regulation of the second DSS coiling. As the line voltage of the traction network is regulated the alternating current side, the valve unit of the traction substation could be uncontrolled, having almost active resistance (irrespective of the switching angles), and rectified voltage ripple considerably lower, than in the regulated valve unit with respect to the net. Minimization of harmonic components, generated by this rectifier and other non-linear consumers, is implemented by the means of a parallel active power compensator (Fig. 2) (Kolb 2005).

The combination of the generation smoothness and jet power consumption with high speed of operation makes the suggested system an effective means of influence on the mode of operational transforming substation of the traction network.

The compensation of inactive power components on the low side (including load balancing) is represented in (Kolb 2005).

4 CONCLUSIONS

By regulating the value and the of the booster voltage vector, the required orientation of the vector of the net current towards the voltage vector can be ensured. If the above mentioned shift equal to zero, the jet power of the capacitive type, equal to the jet power of the open-circuit power transformer, is generated into the net.

By introducing into the system a correcting signal, proportional to the constant voltage deviation of the traction network from the target value, by regulating alternating net voltage, the stabilization of the constant voltage is achieved.

Automatic compensation of the jet power and stabilization of the traction network voltage doesn't depend on external characteristic of the net the load type.

REFERENCES

Brodsky, J., Podaruev, A., Pupynin, V. & Shevelyugin M. 2008. *The stationary energy storage system recovery underground electric rolling on the basis of capacitive energy storage* (in Russian). Electrical Engineering, Issue 7: 38-41.

Brooms, V., Idelchik V. & Liseev M. 1985. *Voltage regulation in electric power systems* (in Russian). Moscow: Energoatomizdat: 247.

Fedorov, A. & Serbinovenogo, G. 1980. *Handbook of electric power supply industry. Industrial Networking* (in Russian). Moscow: Energy: 560.

Zhezhelenko, V. 2000. *The higher harmonics in power systems* (in Russian). 4th ed. rev. and add. Moscow: Energoatomizdat: 311.

Inkov, Y. & Klimash, V. 2003. *Reactive power compensators with the stabilization of the load transformer substations* (in Russian). Electricity, Issue 12: 11 – 16.

Inkov, Y., Klimash, V. & Svetlakov D. 2007. *Compensators inactive power with voltage stabilization transformer substations* (in Russian). Electrical Engineering, Issue 7: 34 – 37.

Kolb, A. 2006. *Power factor correction and stabilization of the output voltage of traction substations* (in Russian). Bulletin of the Dnipropetrovsk National. Univ of Rail. Transport, Issue 10: 14 –17.

Kolb, A. 2005. *Management gating compensators inactive components of the total power* (in Russian). Azov State Technical Bulletin. Univ: Proc. Science. Works, Issue 15, Part 2: 87 – 91.

Energy Efficiency Improvement of Geotechnical Systems – Pivnyak, Beshta & Alekseyev (eds)
© 2013 Taylor & Francis Group, London, ISBN 978-1-138-00126-8

Methods of improving the reliability of distribution networks 6-35 kV

M. Kyrychenko & A. Akulov
State Higher Educational Institution "National Mining University", Dnipropetrovs'k, Ukraine

ABSTRACT: The causes of damage to distribution networks are considered and the main methods for insulation condition monitoring as well as fault detection and location are described.

1 INTRODUCTION

At present, requirements to the reliability and continuity of power supply for industrial enterprises are increasing. The reliability of power supply systems is largely determined by the failure-free operation of transmission lines, and the distribution networks 6-35 kV constitute a considerable part of them. It is known that most faults that occur in the power supply systems (about 80%) fall exactly on the distribution networks. The distribution networks fault analysis shows that up to 60-90% of all failures are the ground faults.

The purpose of this study is to analyse the up-to-date methods used to improve the reliability of power supply systems and electrical safety conditions of the maintenance personnel.

2 BASIC DATA

The main causes of damageability and insufficient security levels of power supply systems, of both service personnel and people and animals contacting with them are (Shkrabets & Kyrychenko 2007):

– imperfection of power supply schemes;

– imperfection of rules of operation and their proper execution;

– absence of protective signaling systems dealing with the single-phase short circuits on the ground;

– absence of diagnostic systems for assessing isolation condition;

– high levels of internal overvoltage;

– use of equipment (switching equipment, cables) which exhausted its regulatory resources.

The single-phase ground short circuits represent the most common type of faults in distribution networks. They are dangerous for both the electrical equipment and the staff due to the peculiarities of network and electrical equipment operation,

The ground fault is an asymmetric type of damage and it is characterized by the appearance of zero sequence components in the network. The voltage and the current zero sequence parameters in the transient and steady-state modes depend on many factors, the principal of which is the ground fault and neutral point operation mode.

The danger of single-phase short circuits is connected with a high voltage impact on the phase insulation, including the emergence of significant overvoltages, which may reach up to 1.73 from the phase voltage value during a zero resistance fault. The vector diagram (Fig. 1) clearly shows the phase voltages distribution for ground short circuits (Shkrabets & Akulov 2010).

By the nature of damage, ground faults are divided into metallic (zero resistance) and arc (across the intermittent arc and across the contact resistance at the points of damage).

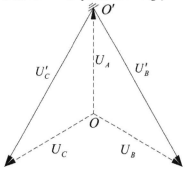

Figure 1. The voltage vector diagram in a single-phase voltage ground fault

Single-phase ground or enclosure faults result from the aging, mechanical damage or electrical breakdown of the insulation of one of the network phases related to the ground or the enclosure. Therefore the task of providing effective control over the insolation condition, as well as early detection and elimination of defects remains topical so far.

The majority of isolation control devices signal that in the power network there is a decrease of resistance and they aren't able to detect the fault location selectively. Sometimes the problem of detecting the fault location in the insulation is solved by the serial electrical separation of the system elements with subsequent monitoring the insulation resistance of a disconnected element. By using this way of detecting a damaged

element there arises a danger of the relay protection and automation malfunction and this requires large expenditures of time and highly qualified personnel (Bulychev, Guliaev, Mishchenko & Shyshygin 2004).

Depending on the line type (cable or overhead) various methods are used for detecting fault locations. Several classifications of these methods are offered in the literature. The most common methods of detecting fault locations for the underground cables are shown in the following diagram (Fig. 2):

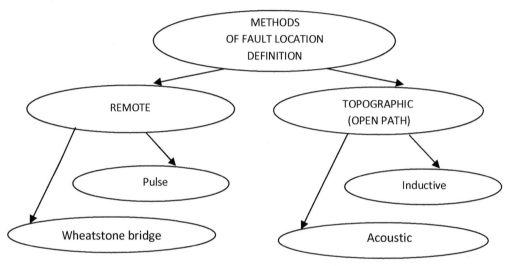

Figure 2. The most common methods of detecting faults of underground cables

The remote methods can be used for the solution of the following tasks:

– measuring the length of cable or overhead lines of communication, power transmission, monitoring, control, etc;

– measuring the distance to the fault or homogeneity location of the line;

– determining the type of line fault (break, short circuit, leakage in the cable insulation, appearance of the additional longitudinal resistance in the conductors, etc);

– measuring the cable line parameters (for example, insulation resistance).

The pulse method is based on the theory of pulse signals distribution across the line. When it is used in for underground cables, the so-called probe electrical pulse is sent and the time of travel of the pulse is measured from the moment it is sent to the moment it comes back being reflected from the fault location. The signal propagation speed depends on the insulation

between conductors. If the line is homogenous and doesn't contain any faults, the pulse signal smoothly travels from the beginning of the line to its end. If there are inhomogeneities on its way (fexample, breakdown of insulation between conductors), a part of the pulse energy passes through this heterogeneity, and another part is reflected and begins to propagate in the opposite direction – to the beginning of the line. In the case when the line is short-circuited or broken, the whole pulse energy is reflected and returned to the beginning of the line. By measuring the delay time of the impulse sent and accepted from the line, it is possible to determine the distance to the fault location.

The Wheatstone bridge technique, based on a direct current or an alternating current of frequency from a few to several hundred hertz, is used to measure the cable insulation resistance, loop resistance (two conductors shorted at the end), cable capacity, distance to fault location, distance to the high-resistance insulation leakage of the line.

The topographical methods are used to detect the required location on the track, that is, topographic points of the fault location.

The induction method is based on the principle of sound recording from the ground surface which is created by electromagnetic fluctuations when the current of sound frequency (800 – 1200 Hertz) is passing through the cable conductors.

The acoustic method is based on recording the sound vibrations over the fault location that occurring in the fault location due to the sparks from electrical impulses that are sent to the cable line.

The greatest efficiency can be achieved by the joint use of both remote and open path methods.

When solving the problem of network protection from the single-phase short circuits, one of the main objectives is to optimize the neutral, as the way of neutral grounding not only determines the operating conditions of network insulation, but also affects the functioning of the automation devices and relay protection, as well as the principles of their construction, on the basis of which, in their turn, concrete structures and schemes of ground fault protection devices are created. The lowest level of operational reliability corresponds to the networks with a fully insulated neutral (Shkrabets, Dvornikov, Ostapchuk & Skosyrev 2003).

Besides, the compensation of capacitive currents of a single-phase ground fault is widely used in order to increase the reliability of electrical networks 6-10 kilo Volt, which allows the current in a fault location to be reduced to the level of an active component and high harmonics, and thus to create conditions for the voluntary liquidation of the network accident.

The experience of operation of electrical networks with automatic compensation of capacitive currents shows that under the single-phase ground faults (SFGF) with currents of more than 100 ampere, the share of faults that pass into the inter-phase short circuits, doesn't exceed 3-5%. Under the SFGF currents of more than 100 ampere, this share increases and at 300-40 ampere and more almost every second single-phase ground fault transfers into the inter-phase short circuits. Therefore, the compensation of only one capacitor component isn't a sufficient condition for the reliable operation of power networks with high single-phase ground fault currents. According to the research data, (Petrov 1975), the average value of a single-phase ground fault current in power networks 6-10 kilo Volt is

125 ampere and in separate power networks (about 5% from the total number of networks) it reaches 400-500 ampere. It should be noted that in connection with the development of cable networks, this, in the long run, leads to an increase of the average value of the SFGF currents. Therefore, to improve the reliability of electric networks, it is necessary to foresee the compensation of an active component and high harmonics of the SFGF residual currents. With the significant value of a single-phase fault current (about 200 ampere and more), the uncompensated current value also increases and there remains a danger of the single-phase circuit transition to the double circuit on the ground or the danger of a short interfacial circuit.

As practice shows, the values of the uncompensated current is enough for the burning of an intermittent arc and for the development of further failure. If to take into consideration that the most significant source of high harmonics are consumers with a non-linear load and their power is constantly growing, there is a need of developing a system of automatic compensation of high harmonics of a short circuit current.

Today, there are two basic principles of active component compensation of the single-phase ground fault current. One of them is based on the creation of the artificial asymmetry in the network and another one – on the introduction of additional voltages into the network. For their implementation a number of statistical devices connected to network elements are known:

– an additional capacity ΔC is connected to the lagging phase;

– inductance ΔL is connected to the anticipatory phase;

– an additional voltage $\dot{U}_{Ad}$, which coincides with the phase voltage of the damaged phase or is ahead of it at a certain angle, is introduced into the neutral through an arc-suppression coil or a single-phase transformer.

The advantage of the passive compensation method is that it can be relatively easily implemented in electric networks of about 1000 Volt. However, its use in networks with the voltage of 6, 10 and 35 kilo Volt is very problematic.

The method, based on the inclusion of an additional source in the neutral of the compensating network, is devoid of this disadvantage; its advantage lies in the simplicity of controlling the input voltage value. Its disadvantage consists in the impact of the compensation devices of an active component on the setting of an arc-suppression reactor.

It is necessary to realize the compensation of the fault current active component under high values of the total ground fault current. Let's consider a scheme of replacement of the three-phase power network, showing only the active component of the process flow in a single-phase ground fault mode (Fig. 3).

Under the single-phase ground fault, on a neutral of the network there appears the voltage $\dot{U}_0$, conditioned by the phase voltages $\dot{U}_A$, $\dot{U}_B$, $\dot{U}_C$ of the network and the inducted voltage $\dot{U}_{Ad}$. The ground fault is equivalent to the inclusion in the fault location of the source, the voltage of which is equal to the damaged phase voltage $\dot{U}_{\hat{O}\bar{\imath}}$ and is opposite in the phase.

After that, in the fault location there appears an active component of the ground fault current $\dot{I}_A$, for the compensation of which the value $-\dot{I}_A$ is used.

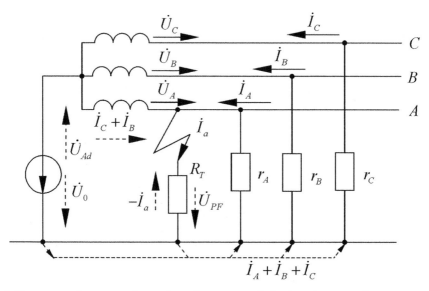

Figure 3. The compensation process scheme of an active component in the three-phase network

The condition of accurate setting of a compensation mode is the equality of neutral voltage $\dot{U}_0$ and faulty phase voltage $\dot{U}_{PF}$. In this case, the only current which will flow through the resistance circuit $R_{\bar{\imath}}$, will be the current, characterized by the presence of high harmonics. To restore the normal operation mode of the compensated network after the emergency shutdown or self-liquidation of the ground fault, it is necessary to reduce the value of the input voltage $\dot{U}_{Ad}$ to zero. This will decrease the neutral voltage $\dot{U}_0$ and the voltages relating to the ground on the damaged and undamaged phases will be restored.

In addition to the main frequency, in the damaged network there appear additional components of high harmonics currents.

With the ground fault, capacities of wires of undamaged phases relative to the ground are switched on the linear voltage. This ensures a high level of harmonics in the ground-fault current passing through the capacities as the capacities are much lower resistance for them than for the current of the main frequency.

In the case of phase-to-ground fault, the ground fault current depends on the equivalent EMF generator, which is equal to the phase voltage in the point of damage in the previous mode. The equivalent circuit of the distribution network is shown in Figure 4:

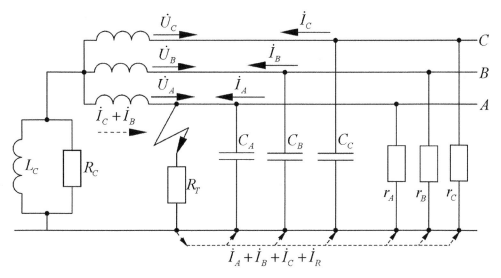

Figure 4. The distribution network circuit with the compensated neutral for single phase ground faults through the contact resistance

The only natural harmonic sources of the ground fault current, which should be taken into account, are the phase voltage distortions in the point of damage under the zero-sequence open-circuit generated by the consumer loads with non-linear characteristics. The natural current of each harmonic of the v-th order that closes in the damage point is:

$$I_{v1} = \frac{U'_{pv1}}{\sqrt{R^2 + \dfrac{1}{3C\omega v^2}}} = \frac{I_c U'_{pv1} v}{U_{pv1}\sqrt{\left(\dfrac{Rv}{X_c}\right)^2 + 1}}$$

where U'_{pv1} – a set of possible voltage harmonics; I'_c – the operating frequency component of the capacitive current line in the absence of damages (in an expression $X_c = \dfrac{1}{3C\omega}$).

In all the cases, when high harmonics of voltage or of current impose on each other they are joined with the component of the basic harmonic squared, that is, the resulting value, obtained as a result of addition, is equal to the square root of the sum of the squares of all the components of different frequencies.

Hence, it follows that even a relatively small harmonic voltage can cause a significant component of the ground fault current. If, for example, a phase voltage contains the ninth harmonic which is equal to 2% of the total

voltage, it produces a corresponding harmonic current equal to 17.8% of the capacitive current of the main frequency. In addition to the considered methods, there are other ways of determining the level of high harmonics in the ground fault current.

For the protection of the distributive network 6-10 kilo Volt with the isolated neutral from the single-phase ground faults (Bulychev, Navolochny & Pozdeev 2004) a switching method is offered. The device, based on its use, makes it possible to continuously monitor the main insulation parameters and in the case of gradual impairment of the latter (for example, when the insulation is aging) it is possible to predict the development of the phase ground faults and at the expense of the released reserve of time to take necessary measures to prevent the single-phase ground faults before they occur. The essence of this method is that in each time point, by means of the switchboard and on the basis of diodes, the phases with the highest and lowest absolute voltage values relative to the ground are selected in the network.

These phases are shunt connected with the ground through restricted resistances. The resulting leakage current, measured by the instantaneous values of shunt voltage, contains the full information on the fault location and the total insulation resistance of the network relative to the ground.

An important integral element of security and reliability of electrical safety systems is the knowledge of the condition of insolation

129

networks and electric installations. All the presently known methods of determining the insulation parameters of electrical installations and networks can be classified as follows (Shkrabets, Tsyplenkov, Kovalev & Kyrychenko 2010):

- using the operating voltage of an electrical installation as a measuring one;
- using the voltage of an external source of industrial frequency as a measuring one;
- using the voltage of an external source of a non-industrial frequency as a measuring one.

In practice, for monitoring of insulation the indirect methods are used, that is, the total insulation resistance of a certain network, related to the ground, is calculated by the value of a single-phase short circuit current. Various methods of measuring the single-phase ground fault currents can be divided into direct and indirect ones and it is recommended to apply indirect methods of measuring the short circuit currents more widely.

The insulation characteristics of electrical networks are directly connected with the values of fault currents and the mode of transient processes under the most common damages – the single-phase ground faults. In this regard, it is possible to state that monitoring of the insulation parameters is associated with the processes:

- of localization (shutdown) of places with damaged insulation – monitoring of the total active insulation resistance in the distribution network;
- of minimization (the capacitive component compensation of the ground fault current and the control over the compensating device setting) of the fault current values – monitoring of the total insulation resistance in the distribution network;
- of optimization of the network neutral modis, that is, management of the quality of the distribution network in order to suppress transient ferroresonance processes which accompany the emergency modes.

The continuous and automatic control of values of electric network insulation components (of the active and capacitive ground insulation resistance of the main phases, the inductance of the compensating device) will allow the emergence of dangerous conditions in the distribution network to be prognosed. The system of continuous insulation monitoring is necessary for the realization of these tasks.

The essence of the method of the continuous and operational control over the electrical network ground insulation and its elements is that the two operational sinusoidal signals are imposed on an electric network relative to the ground, the frequencies of which are not equal and they differ from the commercial ones (Shkrabets, Misyats & Kyrychenko 2008). On the controlled areas (a line or an affixion) and also in the place of operational source connection, the devices are installed the purpose of which is to remove and process the corresponding signals. After the calculation on the basis of a pre-set program the signals corresponding to the insulation values and to the controlled areas of the power supply system are sent.

Thanks to the micro-computer, this method can be used:

– for the operational measurement of the active insulation resistance level of both the whole power network system and each of the affixions of the distributive network;

– for the operational measurement of the related to the ground capacity level of both the whole power network system and each of the affixions of the distributive network;

– for the operational measurement of the arc suppression coil inductance values (compensating devices);

– to carry out selective leakage protection or protection from ground faults in power systems (quarries and mines) regardless of the configuration and mode of operation of the neutral in the network;

– for the automatic setting of a compensating device in resonance with the distributive network capacity.

Many devices of the relay protection and automation use the method of symmetric components. The operation principle of the zero sequence current transformer is based on the addition of current values in all three phases of the protected area. In a normal (symmetric) mode the sum of the phase currents is zero.

When a single-phase short circuit emerges, currents of zero sequence appear in the network and the sum of currents in the three phases won't be zero. This will be fixed by a measuring device (for example, an ammeter) connected to the secondary winding of the zero sequence transformer.

The components of the reverse sequence emerge with the appearance of any asymmetry in a network: a single-phase or two-phase short circuit, phase breakthrough, load asymmetry. The zero-sequence components occur during the ground faults or breakthrough of one phase or two phases. In the case of the phase-to-phase fault the zero sequence components (currents and voltages) are equal to zero.

Each of the considered methods has its advantages and disadvantages.

3 CONCLUSIONS

Improvement of the existing methods as well as the development of the new techniques and tools of monitoring the insulation condition as well as detecting and liquidating the faults in distribution networks will increase the power supply reliability and improve the conditions of indirect energy security in the distribution networks 6-35 kV.

REFERENCES

Shkrabets, F. & Kyrychenko, M. 2007. *Methods for fault location in distribution grids* (in Russian). Mining Electrical Engineering and Automation, Issue 79: 8-13.

Shkrabets, F. & Akulov, A. 2010. Methods *of reducing the total single-phase ground fault current in open-pit mining 6-10 kilo Volt grids* (in Russian). Mining Electrical Engineering and Automation, Issue 84: 83-91.

Bulychev, A., Guliaev, A., Mishchenko, D. & Shyshygin, S. 2004. *The method of insulation monitoring in grids with compensated neutral according to measurements in the transformer neutral* (in Russian). Bulletin of the Vologda State Technical University, Issue 4: 22-25.

Shkrabets, F., Dvornikov, A., Ostapchuk, A. & Skosyrev, V. 2003. *Reliability and electrical safety in distribution grids of mining companies* (in Russian). Mining Informational and Analytical Bulletin, Issue 3: 205-207.

Petrov, O. 1975. *Compensation of active current component at single-phase ground fault in electrical grids* (in Russian). News of higher education institutions. Energy development, Issue 10: 52-59.

Shkrabets, F., Tsyplenkov, D., Kovalev, A. & Kyrychenko, M. 2010. *Insulation parameters of distribution grids: control and management* (in Russian). Electrical Engineering and Electromecanics, Issue 1: 49-51.

Bulychev, A., Navolochny. A. & Pozdeev, N. 2004. *Control of insulation condition in the 6-10 kilo Volt grid with the isolated neutral* (in Russian). Bulletin of the Vologda State Technical University, Issue 4: 13-18.

Shkrabets, F., Misyats, Ye. & Kyrychenko, M. 2008. *The structure of continuous insulation monitoring system in the 6-10 kilo Volt grids in mining enterprises* (in Russian), The Efficiency and Quality of Industrial Enterprises Power Supply: VI International Scientific and Technical Conference. Publishing office PSTU: 401-404.

Energy Efficiency Improvement of Geotechnical Systems – Pivnyak, Beshta & Alekseyev (eds)
© 2013 Taylor & Francis Group, London, ISBN 978-1-138-00126-8

Dynamic objects parameters control on the basis of rebuilt spectral operators application

M. Alekseyev & T. Vysotskaya
State Higher Educational Institution "National Mining University", Dnipropetrovs'k, Ukraine

ABSTRACT: The method of informative features formation for dynamic objects parameters control on the basis of adjustable and adapted to random signals classes' references matrix spectral operators is suggested.

1 INTRODUCTION

Control objects parameters check on the prorated to their functioning signals allows estimating the control object's functional state and detecting the defects beginnings for the purpose of emergency situation prevention measures assuming.

However, the existing systems of parameters control signals disregard the control objects' individual peculiarities. For instance, the value of different supports' absolute compliance 3-6 times differs in one and the same turboset. It means that equal in value driving forces will cause different amplitudes of different supports. In this case vibration check needs individual approach to each turboset bearing vibration check.

2 PROBLEM STATEMENT

The elaboration of the mathematical basis for dynamic objects state estimation constitutes the subject of our paper. This basis is a constituent of such systems states control general problem. The paper deals with the method of informative features formation. These features characterize the control object specific state. They also allow estimating the parameters and classifying the detected defects state and identification.

3 PUBLICATIONS ANALYSIS

The methods based on the orthogonal transformations, especially the ones possessing fast calculating algorithms, should be marked out among widely used informative features singling out methods. Orthogonal transformations, in the case of appropriate basis system choice, secure the adequacy of analyzed information along with the high degree of informative components decorrelation. However, traditional spectral transformations by Fourier, Walsh, Haar (Ahmed & Rao 1980; Alekseev 2000) do not secure the range of low dimension characteristics. From the other hand, the majority of objects with cyclical operation manner functioning is accompanied by diverse physical processes (vibrating, acoustic, etc.) which can be viewed as periodically correlated. It allows estimating each signal class with a proper standard and makes it possible to build the bases adjusted to the standard (Solodovnicov & Spivakovsky 1986).

The approach to rebuilt matrix operators formation based on their concept through a common spectral core ($p \times p = 2 \times 2$), that allows infinite set of functions with fast transform algorithms basis systems obtaining is covered in (Solodovnicov & Spivakovsky 1986). Therewith the spectral operator matrix is presented in a factorized form:

$$H_n = G_n, G_{n-1}, .., G_i, .., G_1 ,\qquad(1)$$

where G_i – rarefactioned by zeroes and further indecomposable matrixes, called Good's matrixes. Each Good's matrix contains 2^{n-1} cores arranged in the following way (for $n=3$):

$$G_r = \begin{bmatrix} [& V_{r1} & &] & & \\ & [& V_{r2} & &] & \\ & & [& V_{r3} & &] \\ & & & [& V_{r4} &] \end{bmatrix}\qquad(2)$$

The core general form is:

$$V_{rl} = \begin{bmatrix} \cos\varphi_{rl} & e^{j\theta_{rl}}\sin\varphi_{rl} \\ \sin\varphi_{rl} & -e^{j\theta_{rl}}\cos\varphi_{rl} \end{bmatrix},\qquad(3)$$

where

$$\varphi_{rl} \in [0, 2\pi]; \theta_{rl} \in [0, 2\pi]; r = \overline{1, n}; l = \overline{1, 2^{n-1}}.$$

4 KEY MATERIAL PRESENTATION

The paper suggests the method of suboptimal according to Karunen-Loyev rebuilt bases possessing fast transform algorithms formation.

Let's express the rebuilt spectral operator's matrix for class m through $\mathbf{B}^m = \left[\mathbf{B}_1^m, \mathbf{B}_2^m, ..., \mathbf{B}_N^m\right]^T$. The condition of orthonormality holds for the series $\mathbf{B}_i^m$ of matrix $\mathbf{B}^m$, i.e.

$$\mathbf{B}_i^{m^T}\mathbf{B}_j^m = \begin{cases} 1, & i = j, \\ 0, & i \neq j. \end{cases} \tag{4}$$

For each $\mathbf{X}^m$ vector, belonging to m class incoming data vectors, we will gain

$$\mathbf{Y}^m = \mathbf{B}^m\mathbf{X}^m, \tag{5}$$

where $\mathbf{X}^{mT} = \left[x_1^m, x_2^m, ..., x_N^m\right]$;

$\mathbf{Y}^{mT} = \left[y_1^m, y_2^m, ..., y_N^m\right]$.

Since for $\mathbf{B}_i^m$ series of $\mathbf{B}^m$ matrix condition (4) holds, then $\mathbf{B}^{mT}\mathbf{B}^m = \mathbf{I}$. Consequently,

$$\mathbf{X}^m = \mathbf{B}^{mT}\mathbf{Y}^m = \left[\mathbf{B}_1^m, \mathbf{B}_2^m, ..., \mathbf{B}_N^m\right]^T \mathbf{Y}. \tag{6}$$

Expression (6) can also be presented in the following way

$$\mathbf{X}^m = y_1^m\mathbf{B}_1^m + y_2^m\mathbf{B}_2^m + ... + \mathbf{B}_N^m y_N^m = \sum_{i=1}^N y_i^m\mathbf{B}_i^m \tag{7}$$

Informative features should represent the vector of incoming data $\mathbf{X}^m$ adequately in order we could gain their minimal number. We use the root-mean-square criterion to estimate the transformation optimality. We gain $\widetilde{\mathbf{X}}^m$ mark of $\mathbf{X}^m$ vector, having represented it in M members in expression (7). Left N-M coordinate y_i^m are substituted with c_i^m constants. Then

$$\widetilde{\mathbf{X}}^m = \sum_{i=1}^M y_i^m\mathbf{B}_i^m + \sum_{i=1}^N c_i^m\mathbf{B}_i^m. \tag{8}$$

The mistake in $\mathbf{X}^m$ vector presentation by its mark $\widetilde{\mathbf{X}}^m$ can be expressed as a mistake vector

$$\Delta\mathbf{X}^m = \mathbf{X}^m - \sum_{i=1}^M y_i^m\mathbf{B}_i^m - \sum_{i=1}^N c_i^m\mathbf{B}_i^m. \tag{9}$$

Taking into account expressions (7) and (9)

$$\Delta\mathbf{X}^m = \sum_{i=M+1}^N (y_i^m - c_i^m)\mathbf{B}_i^m \tag{10}$$

The root-mean-square mistake is defined as follows

$$\sigma^m = E\left\{\left\|\Delta\mathbf{X}^m\right\|^2\right\} = E\left\{\Delta\mathbf{X}^{m^T}\Delta\mathbf{X}^m\right\}. \tag{11}$$

Substitute (10) into (11) and gain

$$\sigma^m = E\left\{\sum_{i=M+1}^N \sum_{j=M+1}^N \left(y_i^m - c_i^m\right)\left(y_j^m - c_j^m\right) \times \right. \\ \left. \times \mathbf{B}_i^{m^T}\mathbf{B}_j^m\right\}. \tag{12}$$

Taking into account (4), the latter expression can be presented as

$$\sigma^m = \sum_{i=M+1}^N E\{(y_i^m - c_i^m)^2\}. \tag{13}$$

Bearing in mind that the values y_i^m are defined by the type of the proper $\mathbf{B}_i^m$ basis vector, the root-mean-square mistake depends on the combination of $\mathbf{B}_i^m$ and c_i^m. To minimize the root-mean-square mistake let us consider the right part (13) and define the expression

$$\frac{\partial}{\partial c_i^m} E\{(y_i^m - c_i^m)^2\} = \tag{14}$$

$$-2[E\{y_i^m\} - c_i^m] = 0$$

(14) yields $c_i^m = E\{y_i^m\}$. Since

$$y_i^m = \mathbf{B}_i^{m^T}\mathbf{X}^m,$$

then $c_i^m = \mathbf{B}_i^{m^T}E\{\mathbf{X}^m\} = \mathbf{B}_i^{m^T}\overline{\mathbf{X}}^m$,

where $\overline{\mathbf{X}}^m = E\{\mathbf{X}^m\}$ - the mathematical expectation of $\mathbf{X}^m$ vector.

Regarding the fact that the difference $(y_i^m - c_i^m)$ is a scalar value, expression (13) can be presented as follows

$$\sigma^m = \sum_{i=M+1}^{N} E\{(y_i^m - c_i^m)(y_i^m - c_i^m)^{T}\}. \qquad (15)$$

Substitute $y_i^m = B_i^{m^T} X^m$ and $c_i^m = B_i^{m^T} \overline{X}^m$ into (15). Yield

$$\sigma^m = \sum_{i=M+1}^{N} B_i^{m^T} E\{(X^m - \overline{X}^m)(X^m - \overline{X}^m)^{T}\} \times \qquad (16)$$
$$\times B_i^m$$

Denote

$$K_x^m = E\{(X^m - \overline{X}^m)(X^m - \overline{X}^m)^{T}\}, \qquad (17)$$

where K_x^m – covariational X^m matrix.
Consequently,

$$\sigma^m = \sum_{i=M+1}^{N} B_i^{m^T} K_x^m B_i^m. \qquad (18)$$

The elements of B_i^m vectors in (18) define the parameters of B^m spectral operator's cores. Since the parameters do not depend on the initial standard in some cores, it is possible to optimize the basis according to the demands of a definite task.
Introduce the function

$$F^m(\varphi_1^m, \varphi_2^m, ..., \varphi_j^m, ..., \varphi_k^m) = \sum_{i=M+1}^{N} B_i^{m^T} K_x^m B_i^m. \qquad (19)$$

where φ_j – spectral operator core parameters, which do not depend on the type of output data; k – the number of spectral operator cores parameters φ_i, which do not depend on the type of output data $k = 1.. N(n/2-1)+1$.
Consequently, the search for suboptimal according to Karunen-Loyev basis is reduced to the search for function $F(\varphi_1, \varphi_2, ..., \varphi_k)$ minimum, i.e. to the experimental problem solving. Therewith, we should take into account that function F in general case can have the length of local minima on the search hyperplane, which stipulates the application of global optimization methods.
In (Alekseev 2007) the defined criterion defines random vectors X as those of class m: if

$A^{ms} = A_p^m - A_p^s > 0$ for all $s = \overline{1, L}(s \ne m)$,

then $X \in \{X^m\}$, where $A_p^m(Y^m) = \sum_{i=1}^{p}(y_i^m)^2$.

The value $A_p^m(Y^m)$ which is the square of X^m vector projection on p dimension subspace can be viewed as the degree of vector decomposition quality in this subspace. The closer $A_p^m(Y^m)$ value to 1 is, the better the quality of decomposition is.

In (Alekseev 2007) decomposition quality is suggested to be estimated by the entropy of energy distribution on basic discrete functions S^m.

$$S_i^m = -\sum_{j=1}^{N}(y_{ij}^m)^2 \log (y_{ij}^m)^2. \qquad (20)$$

Analyzing vectors $X^s, s = \overline{1, L}$, representing the vectors which characterize the state, firstly it is necessary to valuate vector X^s and then to form the system of informative features vectors $Z^m = [z_1^m, z_2^m, ... z_n^m]^T$. The procedure of these features formation consists in vector X^s orthogonal transformation on suboptimal according to Karunen-Loyev systems of discrete orthogonal functions B^m:

$$Z^m = (1/N)B^m X^s, \qquad (21)$$

where B_N^m - suboptimal according to Karunen-Loyev spectral operator for class m square matrix.

5 CONCLUSIONS

The suggested approach of informative features formation in the process of control objects check possesses definite universality, as it is not connected with physical nature of initial features. This approach adequately meets the demands made to the procedure of parameters formation in the process of objects control. It secures the space of low dimension informative parameters gaining, the efficiency of their formation (the applied bases have fast transformation algorithm) and requires relatively plain apparatus or software support.

135

In the prospect of further investigation it will be reasonable to synthesize the method of suboptimal according to Karunen-Loyev rebuilt matrix spectral operators with fast transformation algorithms for the purpose of informative features which are invariant to incoming data arithmetic shift gaining, and to apply the main concepts covered in this article in order to devise the methods of various classes random processes compression.

REFERENCES

Ahmed N. & Rao K. R. 1980.*Orthogonal transformations in digital signals processing*: transl. from English. Edited by I. B. Fomenko. Moscow: Svyaz': 248.

Alekseev M. A. 2000. *About the efficiency of features formation spectral methods in the process of turbosets vibration diagnostics* (in Russian). Vibrations in engineering and technics, Issue 4: 77-80.

Solodovnicov A. I., Spivakovsky A. M. 1986.*The basis of information spectral processing theory and methods.* Leningrad: 272.

Alekseev M. A. 2007. *The criteria of situations classification in technological processes control* (in Russian). Dnipropetrovs'k: NMU edited volume, Issue 29: 172-177.

Energy Efficiency Improvement of Geotechnical Systems – Pivnyak, Beshta & Alekseyev (eds)
© 2013 Taylor & Francis Group, London, ISBN 978-1-138-00126-8

Control automation of shearers in terms of auger gumming criterion

V. Tkachov, A. Bublikov & M. Isakova
State Higher Educational Institution "National Mining University", Dnipropetrovs'k, Ukraine

ABSTRACT: The solution is offered to a topical scientific issue of developing an auger gumming criterion and designing automatic control algorithm for shearer loaders with a view to decrease energy intensity of coal mining. The sliding average ratio of instantaneous power consumption of the cutting drilling tool drive measured with a time shift equal to the quarter period of auger rotation is recommended as the auger gumming criterion. The simulation results of the automatic control system operation of a shearer loader at thin seams have shown the decrease in energy intensity by 40 – 50 % after applying the offered automatic shearer control algorithm.

1 INTRODUCTION

83.2% of the total industrial coal reserves in Ukraine is found in thin flat-lying seams (up to 1.2 m), 88.4% of which (965) are most suitable for cutter-loader coal extraction. Today, the available power of shearer loaders was enhanced for coal extraction in 0.8–1.3 m thick flat-lying seams. For example, shearer loader UKD 400 (project design by "Dongyprougiemash" institute, Ukraine) is supplied with two individual 200 kWt electric motors for each cutting drive and two individual 30 kWt motors for each feeding drive.

At the moment the two means of automation of shearer loader operating mode are used in all automatic systems of control of shearer loaders (Kowal, J., Podsiadło, A., Pluta, J. & Sapiński, B. 2003). These means are chosen by a shearer's operator. One of these two means of automation (means 1) is the stabilization of shearer movement speed (V – const). The shearer's rate of production is set by operator on the basis of specified productivity required. The shearer will work in this mode until power of electric motor of the cutter exceeds its stable value. After that automatic system of control begins to work in accordance with second mean of automation of shearer loader operating mode (means 2). It stabilizes motor power of the cutter ($P = $ max) by means of regulation of shearer feed rate (V – var). When motor power of the cutter is lower that its stable value automatic system of control switches to means 1 again. The shearer's operator can set the means of automation 2, in which case the automatic system of control will work according to this means all the time.

2 FORMULATING THE PROBLEM

To prevent coal-cutting with stone, shearer loaders for thin seams are designed with the auger end effector of the smaller diameter. Since the end effector is influenced by significant loads, reduction of the auger hub diameter is not permitted. This brought about the necessity to reduce the height of auger blade, which resulted in its smaller loading capacity. The conventional algorithm of feeding rate automatic control has been developed taking into account design peculiarities characteristic of shearer loaders whose end effectors have big loading capacity. This algorithm does not take into account shearer's specific energy consumption in spite of the fact that shearer loader is one of the most power-intensive mining machines. In case of thin seams, with the increase of shearer's feeding rate the auger's gumming begins long before the cutting electric drive works in full power. As a result, shearer works in the long-running high level energy intensity mode.

This paper also presents the way to decrease shearer loader's energy intensity by means of the criterion of auger gumming initiation for automatic control system. The fixation of auger gumming initiation for automatic control system will allow to implement automatic control of shearer for thin seams with stabilization of loading capacity of auger end effector and will reduce shearer's specific energy consumption.

3 RESEARCH TOOL

The research was conducted with the help of imitation model "coalface – auger – cutting drive motor" based on the traditional methods of cutter-loader power characteristics calculations (Pozin, E.Z., Melamed, V.Z. & Ton, V.V. 1984), mathematical description of energy transformation in the electrical drive (Starikov, B.Ya., Azarkh, V.L. & Rabinovich, Z.M. 1981), and the results of numerous studies into cutter-loader static dynamics for imitating the load on the end effector (Dokukin, A.V., Krasnikov, Ju.D. & Hurgin, Z.Ja. 1978). The input parameters of the model are design parameters of UKD300 shearer, coal cuttability samples obtained during the trials of UKD300 on 519 longwall of C_5 seam ("Pavlogradskaya" mine, section #7), and mining and geological parameters of the C_5 seam.

The model represented in fig.1 consists of two parts. The drag torque on the auger M_{aug} is simulated with the help of the blocks of first part of the model (HFC, LFC, CC, LCC). The electromechanical energy conversion is simulated with the help of the blocks of second part of the model (TMEMS, EMT).

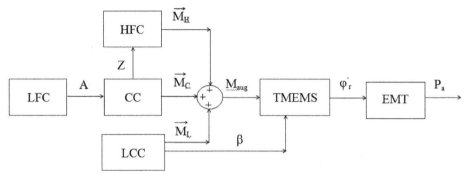

Figure 1. The structure scheme of the imitation model "coalface – auger – cutting drive motor"

The block of imitation of low-frequency change of coal cuttability along coalface (LFC on figure 1) is a generator of random signal, which has the following autocorrelation function:

$$D_A(\tau) = 4241 \cdot 0{,}57 \cdot e^{-\alpha 1 W_f \tau} + 4241 \cdot (1 - 0{,}57) \cdot e^{-\alpha 2 V_c \cdot \tau}, \text{ (N/mm)}^2, \tag{1}$$

where $\alpha 1$ – the attenuation coefficient of correlation function of coal cuttability distribution along coalface, m^{-1}; V_f – the feed rate of shearer loader, m/sec; $\alpha 2$ – the attenuation coefficient of correlation function of coal cuttability distribution along the cutterpath, m^{-1}; V_c – the cutting rate, m/sec; D_A – coal cuttability dispersion, (N/mm)^2.

The output value of block LFC complies with the normal law of distribution with dispersion of 4241 (N/mm)^2 and mathematical expectation of 407 N/mm. The parameters of random value distribution law similarly to the parameters of autocorrelation function (1) are received by means of statistical treatment of samples of coal cuttability obtained during working of shearer loader UKD300 at "Pavlogradskaja" mine.

The current cutting force on the cutters is calculated in the block imitating constructive component of drag torque on the auger (CC, Fig.

1). The method of cutting force calculation is based on the results of experimental investigations of the process of coal massif cutting by an etalon cutter (Pozin, E.Z., Melamed, V.Z. & Ton, V.V. 1984). During these experimental investigations the analytic dependence between cutting force on the cutters Z and coal cuttability A is determined: Z(A). The influence of parameters which differ from etalon values is accounted with the help of appropriate corrective coefficients. The output value of the CC block is the constructive component of drag torque on the auger:

$$M_C(\varphi) = 0{,}5 \cdot D \cdot Z(\varphi, A), \text{ N·m},$$

where D – diameter of the auger, m; φ – the angular displacement of auger, deg.

The block imitating high-frequency component of drag torque on the auger (HFC, Fig. 1) is a generator of random signal, which has the autocorrelation function of white noise. The output random value of HFC block is subordinated to the gamma law of distribution with expectation equaling to zero, and with dispersion:

$$D_H = ((0,56 \cdot Z(A) + 50) \cdot \sqrt{n} \cdot 0,5 \cdot D)^2, (N \cdot m)^2,$$

where n – the quantity of cutters contacting with coal massif.

In the block imitating the component of drag torque, which deals with coal transportation and loading, (LCC, Fig. 1) the mathematical model of coal transportation and loading by auger of small diameter (0,8–1 m) is used (Boiko, N.G. 2002). According to this model, the pressure of coal on auger surfaces has different change patterns in time during auger rotation, depending on ratio of current values of auger working area and volume of coal in this area, and on the status of coal unloading window. It is possible to discriminate four different phases in the process of coal transportation and loading by auger. During the first phase the coal is being transported by blade at limp state, contacting only with blade. The second phase takes place during the interval of auger turning movement, when coal continues to move practically in the unstressed state, however contacting not only with a blade, but also with the auger tube. The third phase corresponds to the auger operating mode when the volume of the coal exceeds the working space of an auger. Therefore the mass of transported coal is being considered as compressed bulk dry medium with consolidated bodies of the cuneiform form. The pressure of coal at this phase increases and considerably exceeds previous phase's values reaching 300 kPa. The further inflow of coal into the auger working space does not change essentially the area of coal contacting with auger's surfaces, which reaches its maximum value. The fourth phase begins at the moment of opening the coal unloading window. Its peculiarity is the simultaneity of two processes happening during this phase of the auger rotation – unlading of coal from the working space and its lading on a face conveyor. The fourth phase, as well as the third one, is characterized by high power consumption during unlading of coal because of the considerable resistance at unlading. This resistance at unlading is conditioned by small area of a unlading window, ledge of conveyer from working side and housing of the drive gear of cutting from non-working side of auger.

For each phase there is determined the analytical dependence of pressure of coal on auger's surfaces on angular displacement of auger $p_c(\varphi)$ as a result of processing data received during numerous experiments. The output value of the LCC block is the component of drag torque, which deals with coal transportation and loading by auger:

$$M_L(\varphi) = D \cdot p_c(\varphi) \cdot S(\varphi) \cdot [f \cdot cos(\gamma) + cos(\gamma)], \, \text{N} \cdot \text{m},$$

where S – the area of coal contacting with auger's surfaces, m^2; f – the coefficient of resistance to coal moving; γ – the angle between the normal to blade surface and cross sectional plane of auger, rad.

The second output value of block LCC is the coefficient of a linearized internal resistance of a mechanical part of the electric drive of cutting:

$$\beta = 2 \cdot \xi \cdot \sqrt{c \cdot J}, \, \text{N} \cdot \text{m} \cdot \text{sec},$$

where J – the joint moment of inertia brought to the electric motor shaft, $kg \cdot m^2$; c – stiffness factor of a transmission line of the electric drive, brought to the motor shaft, $N \cdot m$; ξ – the relative dimensionless coefficient equivalent to viscous friction:

$$\xi = \begin{cases} 0,15 \, , \text{during phase 1} \, ; \\ 0,21 \, , during \, \text{phase 2}; \\ 0,7 \, , during \, \text{phases 3 and 4}. \end{cases}$$

Block TMEMS represents the equations for two-mass analytical electromechanical scheme which describes the electromechanical part of the drive gear of cutting:

$$\begin{cases} M_{turn} - M_{el} = J_r \cdot \varphi''_r; \\ M_{el} - M_{aug} = J_{aug} \cdot \varphi''_{aug}; \\ M_{el} = c \cdot (\varphi_r - \varphi_{aug}) + \beta \cdot (\varphi'_r - \varphi'_{aug}), \end{cases}$$

where M_{turn} – the turning moment of electric motor of cutting drive, $N \cdot m$; M_{el} – the elastic moment arising in the drive gear, $N \cdot m$; M_{aug} – the drag torque on the auger, brought to the motor shaft, $N \cdot m$; φ_r, φ_{aug} – the relative angular motions of motor shaft and auger, rad; J_r – the mechanical moment of inertia of the rotor of the electromotor, $kg \cdot m^2$; J_{aug} – the mechanical moment of inertia of auger, brought to the motor shaft, $kg \cdot m^2$.

In block EMT there is used the mathematical describing the processes taking place at transformation of energy in the induction motor of the drive gear of cutting. It is simultaneous equations of electromechanical transformation of the energy written in the frame of axes α, β. The output value of the model is the active power of motor of cutting drive.

4 RESULTS OF INVESTIGATION

Taking into account energy inefficiency of the existing algorithm of the feeding rate automatic control on thin seams, the authors suggest a way to reduce specific energy consumption in coal mining by developing a new principle of shearer loader automatic control on the basis of auger gumming criterion. The authors recommend to use the sliding average of the cutting drive motor power momentary values ratio as such criterion, the values being measured with the time shift equal to a quarter of the auger rotation period:

$$k_i = \frac{\displaystyle\sum_{x=i-\overline{T}}^{i} \begin{cases} \dfrac{P_{x-0,25\cdot T}}{P_x}, & \text{if } \left[\dfrac{x}{0,25\cdot T}\right] - (2\cdot n + 1); \\[2mm] \dfrac{P_x}{P_{x-0,25\cdot T}}, & \text{if } \left[\dfrac{x}{0,25\cdot T}\right] - (2\cdot n) \end{cases}}{\overline{T}},$$

where $\overline{T}$ – amount of momentary values of the power consumed by the cutting drive motor to achieve the sliding average (313); T – amount of the power momentary values measured during the auger rotation period (96); n – some integral number; i – the number of the current momentary value of the numerical criterion k_i; P_x – averaged momentary value of the cutting drive motor power, kWt; $P_{x-0.25\cdot T}$ – momentary power value shifted in time for the quarter of auger rotation period against power value P_x, kWt.

The suggested statistic power assessment allows to fixate the auger gumming because at the beginning of gumming and under certain angular intervals of auger rotation, coal is loaded and transported with the increased coal mass pressure on the blades (areas II and IV, Fig. 2), which results in the growth of power measured within these angular intervals, whereas it does not change significantly during other intervals of the auger rotation period.

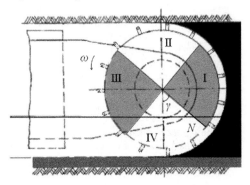

Figure 2. Design scheme of the auger (view from the unloading end side); line N shows the position of the end part of the blade ($\gamma = 50^0$)

Let us investigate the dynamics of numerical criterion of auger gumming without auger gumming at the stationary value of the feed rate and in transient working mode of a shearer loader. For this purpose the imitation model of automatic system of control of shearer loader is developed on the base of the imitation model "coalface – auger – cutting drive motor" (Fig. 3). The imitation model of control system of feed drive of shearer loader is added, which consists of blocks: "Shearer", TMEMS$_{fd}$, EMT$_{fd}$, FC, "PI-regulator", "Block of control" (Fig. 3). In the "Shearer" block the drag torque on the drives of feed drive M_{dr} is calculated with accounting of feeding force Y, which is formed on the auger and is determined in the CC block, and friction force between supporting mechanisms of shearer and underconveyor plate.

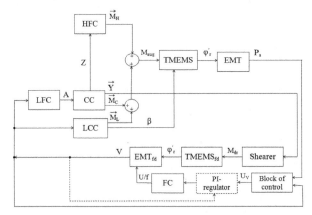

Figure 3. The structure scheme of the imitation model of automatic system of control of shearer loader

Block TMEMS$_{fd}$ is the equations for two-mass analytical electromechanical scheme, which describes the electromechanical part of the drive gear of feeding. The output value of the TMEMS$_{fd}$ block is rotational speed of rotor of electric motor of feed drive φ'_r. In the EMT$_{fd}$ block there is used the mathematical description of the processes taking place at transformation of energy in the induction motor of the drive gear of feeding. The close-loop automatic control of shearer's feed rate V is organized with the help of regulation of amplitude/frequency ratio (U/f) of supply voltage of feed drive electric motor, which is realized in the model of frequency converter (FC block, Fig. 3),. For automatic control of shearer's feed rate the proportional integral regulator ("PI-regulator" block, Fig. 3) is used. The functional algorithm of feed rate control is realized in the "Block of control" at the Figure 3, the output value of which is the set point of shearer's feed rate U$_V$ for PI-regulator.

The dynamics of criterion of auger gumming without gumming is shown at Figure 4. Let us accept the initial moment of sampling the engine power (line N, Fig. 2), when the end of one of blades goes out of the area of the accumulated coal near the unloading side of the auger.

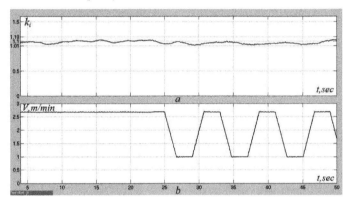

Figure 4. The dynamics: a – of numerical criterion of auger gumming; b – of feed rate of the shearer loader

As we can see from Figure 4, both at a constant value of a feed rate, and at transient mode of a shearer loader, there is observed practically invariable value of numerical criterion of auger gumming. To be more precise, there are observed random oscillations of numerical criterion k_i in a constant range of changing of values of 0.99–1.21. The random oscillations of numerical criterion occur because of limited time of averaging (effect of high-frequency component of engine power is not completely compensated) and because high-frequency oscillations of engine power are modulated by low-frequency ones (effect of low-frequency component of engine power is not completely compensated).

Let us investigate the dynamics of numerical criterion of auger gumming in case of gumming.

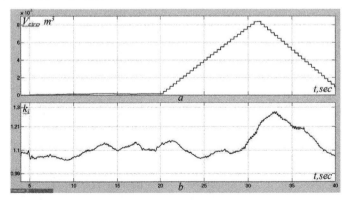

Figure 5. The dynamics: a – of the volume of coal circulating in working space of auger; b – of the numerical criterion of auger gumming

141

From Figure 5 we can see that the numerical criterion of auger gumming is being changed randomly within the range of values 0.99–1.21 with the volume of circulating coal V_{circ} less than value of $8.2 \cdot 10^{-3}$ m^3 (33% of the auger working space) and goes out of the given range in case of exceeding the volume of circulating coal with the value of $8.2 \cdot 10^{-3}$ m^3. During research it was established that the volume of circulating coal at which the numerical criterion exceeds 1.21 is an aleatory variable. It is explained by random oscillations of numerical criterion because of the influence of engine power component of the cutting drive, which is associated with coal cutting. However, if after numerical criterion goes beyond the range of change of values 0.99– 1.21 we shall decrease the feed rate to eliminate circulating coal (see Fig. 5), the numerical criterion will return into a range of changes of values 0.99–1.21 at decrease of volume of circulating coal down to 22– 26 % of the working volume of the auger, and will become less than average value of 1.1 at volume of circulating coal 6.8–10.5 % of the working volume of the auger. The given regularity of dynamics of numerical criterion can be used in the mode of tracking critical feed rate over gumming to determine the moment when circulating coal is practically removed from working space of the auger and when it is possible to again increase the feed rate for the iteration step.

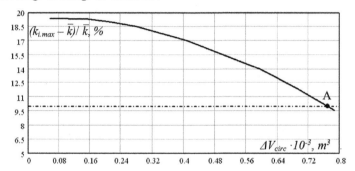

Figure 6. The relationship between a maximum relative deflection of numerical criterion from mean value at gumming and an increase of volume of circulating coal during auger rotation

To develop the algorithm of feed rate automatic control it is necessary to substantiate the critical maximum value of the increase of volume of circulating coal during auger rotation. With this purpose the graphical there is obtained the relationship between maximum relative deflection of numerical criterion from mean value at gumming and the increase of volume of circulating coal during auger rotation (Figure 6).

So as the losses of productivity of a shearer loader during determination of critical feed rate over gumming were minimal, it is necessary to consider the increase of volume of circulating coal during auger rotation equal to the value ($\Delta V_{circ.crit}$), when the relative deflection of numerical criterion exceeds 10% (see point A, Fig. 6). This value equals to $0.76 \cdot 10^{-3}$ m^3 (3.1 % of the working volume of the auger). It follows herefrom that for fixation of gumming of the auger by the automatic system of control of shearer loader with probability "1" in conditions of computational experiment, it is necessary that the increase of volume of circulating coal during auger rotation did not exceed value of $0.76 \cdot 10^{-3}$ m^3 (3.1 % of the working volume of the auger).

The development of the algorithm for the shearer loader's automatic control for thin seams was based on the obtained results of the research into the auger gumming criterion change in time to reduce specific energy consumption in coal mining. According to this algorithm, it is possible to define two working modes of shearer loader's automatic control system – mode of determining the feed rate that is critical in terms of gumming and the following mode. In the search mode, the shearer loader's automatic control system discretely increases the feed rate at intervals determined according to the maximum acceptable value of the circulation coal volume growth per auger rotation (3.1% of the auger working space) and the intervals defined by the time required for the gumming criterion analysis. In the following mode the automatic system periodically switches the shearer loader into the auger gumming operation mode and checks the value of feed rate critical in terms of gumming to ensure the shearer loader works with maximum loading capacity.

Let us imitate the work of the automatic system of control of shearer loader in accordance with the existing algorithm of control of feed rate described in the introduction.

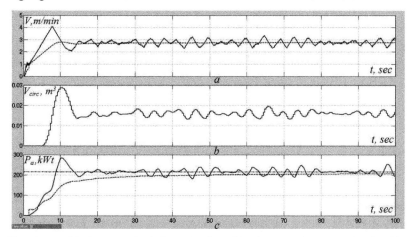

Figure 7. Time change of feed rate (a), circulation coal volume (b) and mean active power of cutting drive motor (c) under automatic control of shearer loader according to the existing algorithm

The instantaneous values of parameters at Figure 7 are represented by solid lines. The results of averaging of these parameters are represented by dotted lines.

According to the mode of cutting drive motor power stabilization the automated control system (ACS) increases feed rate to the moment when the active power of the cutting drive motor exceeds its stable value (215 kWt, see Fig. 7,c). ACS of the shearer loader will keep increasing feed rate despite auger gumming, which starts at the exceeding of the loading capacity of the auger (see Fig. 7,b). When the active power of the cutting drive motor exceeds its stable value, ACS of the shearer loader begins to reduce the feed rate (see Fig. 7,a) unless the circulation coal pressure on the surfaces of the auger is compensated by the power supply of the cutting drive motor in conditions of the shearer loader operation on the edge of auger gumming. Then ACS stabilizes the active power of the cutting drive motor at the stable value level (see Fig. 7,c). During this stage one can see circulation coal occupying about 64% of the auger working space (see Fig. 7,b).

Thus, the shearer loader operation under the feed rate automatic control according to the existing algorithm takes up maximum cutting drive motor power and significantly increases specific energy consumption in coal mining. This is accounted for by the fact that 42% of the cutting drive motor power is used to perform ineffective work – overcoming pressure of the circulation coal and its force of traction against the auger surface.

Modeling of ACS work in accordance with the suggested algorithm of control of feed rate shows that the feed rate changes step by step with slight deviations from the average value which corresponds to the maximum auger loading capacity (see Fig. 8,a). The circulation coal volume in the auger working space starts growing as feed rate increases, and goes down as the feed rate decreases (see Fig. 8,b). Its maximum value varies within the range of 31–35 % of the auger working space, which is 2.9 times lower than the value under the existing algorithm of control of feed rate.

143

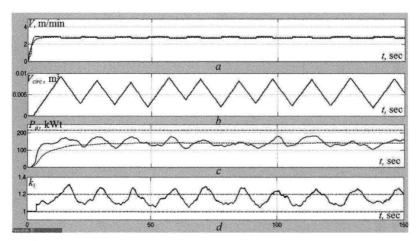

Figure 8. Time change of feed rate (a), circulation coal volume (b), mean active power of cutting drive motor (c) and auger gumming criterion (d) under automatic control of shearer loader according to the suggested algorithm

The boundaries of the range within which the cutting drive motor power varies have significantly diminished, because of the 2.9 times lower average circulation coal volume in the auger working space (see Fig. 8,c). In conditions of automatic control under the suggested algorithm the power randomly varies from 100 to 190 kWt with the average value of 140 kWt. Under the existing algorithm of the feed rate automatic control the cutting drive motor power becomes stable at 214 kWt. The shearer loader efficiency is the same in both cases and is related to the auger loading capacity. Thus, replacing the existing algorithm of the feed rate automatic control by the suggested one will allow to reduce specific energy consumption in coal mining by 58%.

5 CONCLUSIONS

1. The existing algorithm of the feed rate automatic control for thin seams is energy inefficient because of the constant presence of circulation coal which on average occupies 64% of the auger working space. It takes about 42% of the cutting drive motor power (92 kWt for UKD 300) to overcome the force of coal traction against blade surface.

2. Reduction of specific energy consumption in coal mining is possible by developing a method to fixate the gumming start by the automatic system on the basis of the statistic analysis of the cutting drive motor power. It is feasible to use the sliding average of the cutting drive motor power momentary values ratio as the auger gumming criterion for the values measured with the time shift. The auger gumming criterion changes in time sporadically around its average value within the range of ±10%, if the circulation coal volume is less than 33% of the auger working space; and goes beyond this range, if the circulation coal volume is more than 33% of the auger working space.

3. Modeling of the work of automatic system of control of shearer loader demonstrated that the shift from the existing algorithm of the feed rate automatic control to the suggested one at the expense of significant reduction of the circulation coal volume in the auger working space allows to bring down specific energy consumption in coal mining by 58% (0.413 kWt·h/ton for UKD 300).

Thus, the proposed algorithm of the feed rate automatic control for thin seams is significantly more energy efficient.

REFERENCES

Kowal, J., Podsiadlo, A., Pluta, J. & Sapinski, B. 2003. *Control systems for multiple tool heads for rock mining*. Acta montanistica slovaca rocnik, Vol.8: 162-167.

Pozin, E.Z., Melamed, V.Z. & Ton, V.V. 1984. *Destruction of coal by stoping machines* (in Russian). Moscow: Nedra: 288.

Starikov, B.Ya., Azarkh, V.L. & Rabinovich, Z.M. 1981. *Asynchronous drive of shearers* (in Russian). Moscow: Nedra: 288.

Dokukin, A.V., Krasnikov, Ju.D. & Hurgin, Z.Ja. 1978. *Statistical dynamics of mining machines* (in Russian). Moscow: Mashinostroenie: 239.

Boiko, N.G. 2002. *Coal loading by shearers* (in Russian). Donetsk: DonNTU: 157.

144

Energy Efficiency Improvement of Geotechnical Systems – Pivnyak, Beshta & Alekseyev (eds)
© 2013 Taylor & Francis Group, London, ISBN 978-1-138-00126-8

The decision support system in selecting further industrial enterprise direction

L. Koriashkina
Oles Honchar Dnipropetrovs'k National University, Dnipropetrovs'k, Ukraine

Y. Nikiforova, A. Pavlova
State higher educational institution «National mining university», Dnipropetrovs'k, Ukraine

ABSTRACT: The article presents a decision support system (DSS) that serves for selecting the direction of industrial enterprise development. The DSS is based on the analysis of its internal capabilities and external conditions defined by the world globalization processes. Inductive methods of mathematical modeling were applied to evaluate the extent of the country's involvement in globalization processes. To identify external parameters, evaluate a company's internal situation and provide recommendations for the enterprise expedient development, a fuzzy logic system has been applied.

1 INTRODUCTION

The most expressive feature of the current epoch is the rapid growth of the world globalization processes. Since the trends of escalating globalization have both positive and negative aspects, it is an important task for any competitive operating industrial enterprise, while planning its activities, to take into account possible external conducive conditions or barriers, including those caused by globalization, and as a result –adaptation to these conditions.

The paper describes a decision support system for selecting the enterprise development strategy based on a synthesis of inductive modeling methods and fuzzy logic. In addition, we consider problems associated with implementation of the recommendations prescribed by DSS for particular mining (industrial) enterprises, either unprofitable or successfully functioning in the relevant market.

2 DEVELOPMENT OF THE DSS

The developed decision support system is intended for administration of large companies to plan major complex programs, substantiate decisions concerning inclusion in the program of various activities and distribution of resources between areas of development according to their contribution to the achievement of the main program goals.

DSS is characterized by the following functional capabilities:

— level of the country's involvement in the processes of world globalization is projected on the basis of information about the values of the country's macro-economic factors such as inflation, unemployment, real GDP growth, the country's share in global exports , imports, the index of internationalization, transnationality index etc.;

— external conditions of enterprise operation(the level of risk, competition, access to resource and sales markets, inflation) are defined on the basis of information about the level of the country involvement in the process of globalization;

— the vector of further enterprise development for the near future (3 – 5 years) is determined considering information about external conditions and internal capabilities of the enterprise.

Further we present a more detailed process of developing DSS and mathematical apparatus lying at its base. Separate system elements are described in (Koriashkina 2012; Nikiforova 2012; Galushko 2012).

The economic mathematical model for evaluating the extent of a particular country involvement in the processes of the world globalization is obtained using Group Method of Data Handling (Ivakhnenko1994), which restores functional relations in the form of Kolmogorov - Gabor's generalized polynomial (1):

$$Y = a_0 + \sum_{i=1}^{N} a_i x_i + \sum_{j=1}^{N} \sum_{i \le j} a_{ij} x_i x_j +$$
$$+ \sum_{i=1}^{N} \sum_{j \le i} \sum_{k \le j} a_{ijk} x_i x_j x_k + \dots \qquad (1)$$

where Y – the resulting parameter;

a_i, a_{ij}, a_{ijk} – unknown coefficients;

x_1, x_2, ..., x_n – original settings.

The criterion of regularity (2):

$$\overline{\varepsilon^2} = \frac{1}{N} \cdot \sum_{i=1}^{N} (y_i - f(x_1^i, \dots, x_n^i))^2 \to \min \qquad (2)$$

was applied while constructing a model, where

$\overline{\varepsilon^2}$ –the criterion of regularity;

N –the number of interpolation points;

y_i – set points of the resulting parameter;

$f(x_1^i, \dots, x_n^i)$ –the function obtained from the initial parameters.

The constructed functional relation from the parameter that determines the extent of the country's involvement in the process of globalization (rank position in the world, the index of globalization level) allows to select the most significant macroeconomic indicators. These factors include inflation, unemployment, real GDP growth, market capitalization of the companies on the stock exchange (% of GDP), the country's share in global exports , imports, index of internationalization; transnationality index, the ratio of foreign direct investment to GDP, the index ratio of industrial products to goods trade (% of GDP).

Moreover the received functional relation allows to investigate the trends of globalization by changing one or two indicators over a period of time. The fuzzy inference system for evaluation of enterprise's external conditions is designed using Fuzzy Logic in (Nikiforova 2013). The system has roughly 140 fuzzy products based on research into the world globalization tendencies and macroeconomic factors of the country as well as additional economic information from literature.

Fuzzy inference system has two elements. The first block of the system serves to determine the extent of country's involvement in the processes of the world globalization, namely the country's place in the ranking of economic globalization, on the basis of the values of external economic factors selected by using the model (1) in. On the grounds of this result, are evaluated the environmental conditions in which the enterprise

operates in the second block of fuzzy output system, such as:

— the level of risk;
— the level of competition;
— access to resource markets;
— access to sales markets;
— inflation growth (decline).

The second block of DSS allows determining expedient vector of enterprise development on the basis of the above mentioned external conditions of its functioning, as well as considering its internal state (liquidity, profitability, etc.). Mathematical model of the vector of industrial enterprise development is the following (3):

$$V = F\left(\beta; \gamma_1, \gamma_2, \gamma_3, \dots \gamma_n\right), \qquad (3)$$

where – $V = (\alpha_1, \alpha_2, \alpha_3, \dots, \alpha_m)$ the vector of the industrial enterprise development, which consists of the vectors α_i, $i = \overline{1, m}$ the fuzzy linguistic variables $\alpha_{ij} \in \tilde{A}_{ij}$, $\forall j = \overline{1, n_i}$, each of which specifies a certain development direction: $\alpha_i = (\alpha_{i1}, \alpha_{i2}, \dots, \alpha_{in_i})$ – one of the $i = \overline{1, n_i}$ enterprise development plans (see Fig. 1);

F – a vector function describing the dependence of components of the expedient development vector from external and internal factors (defined by rules of fuzzy production);

β – a vector of four fuzzy linguistic variables that define the enterprise functioning external conditions caused by the world globalization processes (the level of risk, the level of competition, access to the sales markets, access to the markets of resources and inflation). It should be noted that the values of these parameters are defined using the fuzzy inference system developed in the second block;

γ_1, γ_2, γ_3, ... γ_n – are values that describe the conditions of the internal enterprise environment for the current period and are determined by experts.

Each linguistic variable α_{ij} is associated with a fuzzy set (4) of its values $\alpha_{ij} \in \tilde{A}_{ij}$, where

$$\tilde{A}_{ij} = \left\{ \begin{array}{l} \left(a_{ij}^1, \mu_{\tilde{A}_{ij}}\left(a_{ij}^1\right)\right); \left(a_{ij}^2, \mu_{\tilde{A}_{ij}}\left(a_{ij}^2\right)\right); \dots; \\ \left(a_{ij}^k, \mu_{\tilde{A}_{ij}}\left(a_{ij}^k\right)\right) \end{array} \right\}, \qquad (4)$$

$\forall i = \overline{1, m}; \ \forall j = \overline{1, n_i},$

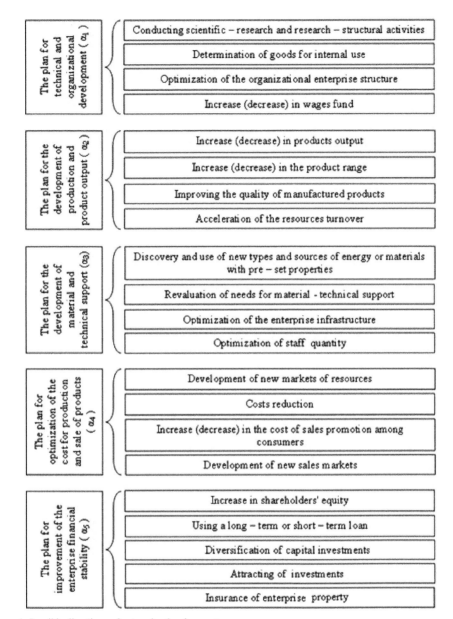

Figure 1. Possible directions of enterprise development

The extent of confidence in choosing a particular direction of enterprise development is determined by the formula (5):

$$\mu_{\tilde{A}_{ij}}\left(a_{ij}^l\right) = \max_{q=1,\ldots Q_{ij}^l}\left(\omega_q^{a_{ij}^l}\prod_{p=1}^{P_q}\mu_{\tilde{B}_p}\left(\beta_p\right)\prod_{s=1}^{S_q}\mu_{\tilde{G}_s}\left(\gamma_s\right)\right)$$

$$\forall i = \overline{1,m};\ \forall j = \overline{1,n_i};\ \forall l = \overline{1,k}\ , \tag{5}$$

where Q_{ij}^l – the number of active production rules that matches the value of the resulting linguistic variable;

$\omega_q^{a_{ij}^l}$ – the weight of the appropriate fuzzy rules in the knowledge base (the number in the spacing [0,1] which characterize the extent of expert's confidence in the virtue of a particular rule);

147

$\mu_{\tilde{B}_p}\left(\beta^p\right)$ – the value of the membership function of input variable β_p to linguistic term from a set $\tilde{B}_p$, $p = \overline{1, P_q}$, P_q – the number of input variables β_p involved in the relevant rule;

$\mu_{\tilde{G}_s}\left(\gamma_s\right)$ – the value of the membership function of input variable γ_s to linguistic term from a set $\tilde{G}_s$, $s = \overline{1, S_q}$, S_q – the number of input variables γ_s involved in the relevant rule.

So, to determine the expedient vector of enterprise development direction, we solved the problem of finding such values of membership functions $\mu_{\tilde{A}_{ij}}\left(a_{ij}^l\right)$, $\forall i = \overline{1, m}$; $\forall j = \overline{1, n_i}$, $\forall l = \overline{1, k}$ using fuzzy (or inexact) information on the quantities β and γ_1, γ_2, γ_3, $\dots \gamma_n$ under which we got the maximum value of (3), i.e.:

$$V^* = \max F\left(\beta; \gamma_1, \gamma_2, \gamma_3, \dots \gamma_n\right), \qquad (6)$$

where V^* – the vector of the industrial enterprise expedient development on the turn of new planning that ensures adaptation to the world globalization and maximum use of the enterprise's opportunities.

Thus, the initial data for the second unit of DSS are evaluations of environmental factors received using the first block, and the indicators of the financial condition of the company:

— the product life cycle;
— the organization life cycle;
— the current liquidity;
— the return on assets;
— the return on equity;
— the turnover of fixed assets;
— financial stability.

It should be noted that in contrast to (Galushko 2012), the rate of inflation is excluded from the external indicators, since the effect of this factor always has an impact on the internal parameters of the enterprise. It is indicated by the formulas that underlie calculation of each of those parameters. In addition, the list of development directions was restructured and changed: first of all, the number of development plans were reduced to five, secondly, directions, carrying a similar meaning, were combined, thirdly, new trends, such as increasing costs of sales promotion, diversification of investments and others were added (Fig. 1).

3 RESULTS OF DSS IMPLEMENTATION

It is rather difficult for many enterprises to remain commercially viable and bring any profit in the current market economy. Approbation of the developed DSS at various Ukrainian enterprises revealed a certain regularity. For instance, if the company is unprofitable, then, as a rule, it is offered to follow the directions that allow getting out of the crisis, such as attraction of long-term loans, optimization of material and technical support etc. Other features are presented in Fig. 2.

So, one of the main results of DSS is the recommendation to make a choice of the optimal development vector based on the fact whether the enterprise is profitable, or at the stage of functioning, it becomes a loss operation. In case the enterprise is unprofitable, before proceeding to the selection of areas for further work it is necessary to provide access to the break-even point by involving a long-term loan. Selection of the development direction scheme considering the financial state of the enterprise is presented in Figure 2. It is obvious that the implementation of the recommended directions presents solution of certain optimization tasks. A mathematical model of one of such tasks will be presented below.

Optimization task for unprofitable enterprises can be formulated as follows: to determine minimum funds necessary for the enterprise to borrow for a certain period (even maybe identify this period) and to relocate these funds to other areas so that the enterprise could get out of crisis by the specified date.

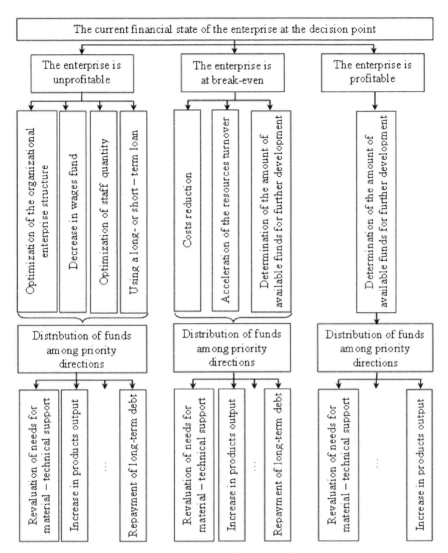

Figure 2. Recommended development directions and the problems related

The mathematical model of the task has the form (7). Enterprise "recovery" indicates the accumulation of a certain amount C_{min} for the possibility of imbedding it in manufacturing or more profitable directions. The obtained funds are proposed to distribute among m directions.

Let us assume that the enterprise can reach the break – even point in k years. To do so, it will need certain amount of money in C_{min} UAH. Then the task will be the following:

$$K \to \min$$
$$x_{10} + x_{20} + ... + x_{m0} = K + V_0$$
$$x_{11} + x_{21} + ... + x_{m1} = \sum_{i=1}^{m} p_{i0} x_{i0} - p_{(m+1)0} K + V_1$$
$$...$$
$$x_{1(k-1)} + x_{2(k-1)} + ... + x_{m(k-1)} =$$
$$= \sum_{i=1}^{m} p_{i(k-2)} x_{i(k-2)} - p_{(m+1)(k-2)} K + V_{(k-1)}$$
$$x_{1k} + x_{2k} + ... + x_{mk} = C_{min} - p_{(m+1)(k-1)} K + V_k$$
$$x_{ij} \geq 0, \quad i = \overline{1,m}, \quad j = \overline{1,k}$$
$$0 \leq K \leq C_{max}$$

(7)

where K – the size of involved loan;

x_{ij} – the amount of funds invested in the ith direction in j^{th} year;

V_j – money received from the realization of other directions;

p_{1j}, p_{2j}, ..., p_{mj} – the share of invested money in the appropriate direction which is returned at the end of jth year;

$p_{(m+1)j}$ – the percentage rate of the loan, expressed as a fraction of units in the jth year;

C_{min} – the size of required funds for getting to the break-even point;

C_{max} – maximum size of loan.

The resulting economic and mathematical model is a linear programming problem which is easy to implement.

When after k years the enterprise makes a profit and is able to move in new directions,the internal state of the enterprise and the external conditions ofits functioning should be revaluatedfor choosing the most suitable vector for further development and the directions clarified using the developed decision support system.

4 CONCLUSIONS

The article presents a decision support system for selecting the enterprise direction, developed from the synthesis of methods of inductive modeling and fuzzy logic. In addition, the issues of implementation of the directions recommended by DSS for unprofitable or successful operating on the existing market enterprises were considered.

REFERENCES

Koriashkina, L., Nikiforova, Y. & Pavlova, A. 2012. *System analysis of globalization processes and their impact on macroeconomic country indicators* (in Russian). Herson: HNTU: Intellectual systems for decision making and problems of computational intelligence.

Nikiforova, Y. 2013. *Planning of industrial enterprise development taking into account the level of the country's participation in globalization processes.* Dnipropetrovs'k: NMU: Scientific Journal of the National Mining University, Issue 3.

Galushko, O., Nikiforova, Y. & Koriashkina, L. 2012. *Choosing effective development directions of industrial enterprises in the context of globalization based on economic and mathematical modeling* (in Russian).

Dnipropetrovs'k: NMU: Economic Journal of the National Mining University, Issue 3.

Ivakhnenko, A. & Madala, H. 1994. *Inductive learning algorithms for complex systems modeling.* London: CRC Press: 368.

Technology mapping of thermal anomalies in the city of Dnipropetrovs'k, Ukraine, with application of multispectral sensors

B. Busygin & I. Garkusha
State Higher Educational Institution "National Mining University", Dnipropetrovs'k, Ukraine

ABSTRACT: Basic results of digital mapping of heat flows arrangement in the city of Dnipropetrovs'k (Ukraine) have been studied taking into account multispectral sensors Landsat TM/ETM+ and Terra MODIS data.

1 INTRODUCTION

With advent of multispectral survey technique fit to take up thermal subband of E-M spectrum (TIR, Thermal Infrared) with wave lengths 8 to 15 µm, investigations of land surface heat flows pattern arrangement and development of respective digital LST maps, gained widespread popularity. LST maps are often used in joint analysis with vegetal indices arrangement maps aimed at identification of suppressed vegetation. Joint application of data on temperature of mapped surface and those from near and middle IR subbands allows for revelation of various thermal anomalies, as well as for reasoning about forest fires location, thermal behavior of industrial wastes lodgement (terricones), leakages in heat supply networks, etc.

Application of LST maps has been given much attention over recent years in investigation of UHI – Urban Heat Island. The reason for that is the importance of assessment of how zones of major thermal anomalies impact both environment and humans.

Spatially, UHI is a model of arrangement of thermal surface field within boundaries of a town, which is significantly higher (3-4 ^{0}C) than that of background outside. Analysis of LST for heavily industrialized territories shows that there may exist comprehensive thermal anomalies inside town limits, characteristic of, say, industrial wastes discharge, production facilities, open ground construction sites etc.

The problem of creation and analysis of LST maps has been much publicized. In publications (Walawender & Haito 2009; Oguro & Tsuchiya 2011) comparative characteristics of the main satellite systems apparatuses are quoted, with the help of which the maps are charted. Some articles (Walawender & Haito 2009; Oguro & Tsuchiya 2011; Qin, Karnieli & Berliner 2001; Nichol J. 2005; Barsi et al. 2003) provide analysis of problems which typically challenge scientists in construction of LST. Among most substantive is the task of assessment of distinctions in temperature values, obtained from different satellite systems.

This paper is aimed at exposition of both method and results of Dnipropetrovs'k thermal field evaluation based on multispectral survey data taken at various times.

2 INITIAL DATA

Exploration target is a 600 км^2 land plot embracing the city of Dnipropetrovs'k and vicinities (Fig. 1). 350 км^2 of area total are attributed to urban and industrial structures, and over 70 км^2 – aquatic area. The remaining part is attributed to agricultural land.

Ten archive satellite images provided input data for research, 5 of these had been obtained with sensor Landsat-5 TM (processing level 1T), the other 5 – with sensor Terra MODIS (processing level 2). Dates and time of imaging are given in Table 1.

3 ESTIMATING PROCEDURES

Let us bring to view procedure of LST mapping on platform of Landsat TM/ETM+ data.

1. Revaluation of thermal channel pixel values (DN), as well as pursuant to points 4 and 5 of Red and NIR channels, into spectral radiance units.

2. Atmospheric correction.

3. Assessment of brightness at-sensor temperature using sensor thermal channel data.

4. Evaluation of mapped surface reflectance.

5. Calculation of normalized difference vegetation index (NDVI) for determination of surface spectral radiation factor.

6. Estimation of surface spectral radiation factor values.

7. Estimation of LST values with application of methods like mono-window, split-window (for data from two thermal channels), as well as technique set out in works (Artis & Carnahan 1982; Nichol 1994, 2005; Xiong et al. 2012; Weng 2001; Weng, Lu & Schubring 2004; Wan 1994).

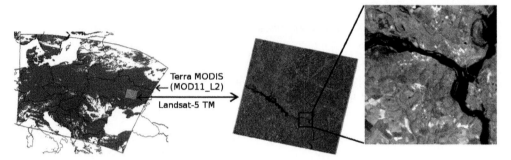

Figure 1. The research territory.

Table 1. Dates and times of surveying.

Num.	Scanner	Date of surveying	Time of surveying (UTC)
1	Landsat-5 TM	22.04.2011	08:20
2		18.06.2011	08:14
3		27.07.2011	08:20
4		28.08.2011	08:19
5		22.09.2011	08:13
6	Terra MODIS	22.04.2011	9:35
7		18.06.2011	9:30
8		27.07.2011	9:35
9		28.08.2011	9:35
10		22.09.2011	9:30

For revaluation of thermal channel pixel values into spectral radiance units one of the two methods is used. Choice is upon availability and character of initial data used for revaluation.

In the first method mentioned, expression (1) is used with DN value of channel picture pixel, as well as gain and bias factors of imaging sensor (Landsat 2008):

$$L_\lambda = DN \cdot gain + bias , \qquad (1)$$

where L_λ – amount of radiation incoming to sensor, $W/(m^2 \cdot ster \cdot \mu m)$; gain and bias calibration values can be found either in metadata files accompanying satellite images, or, alternatively, they are quoted in user's documents.

In the second method the following equation is used (Landsat 2008):

$$L_\lambda = \frac{L_{max\lambda} - L_{min\lambda}}{Q_{calmax} - Q_{calmin}} \cdot (Q_{cal} - Q_{calmin}) + L_{min\lambda} , \qquad (2)$$

where $L_{min\lambda}$ – amount of radiation incoming to sensor, which, after scaling, is converted into Q_{calmin} , $W/(m^2 \cdot ster \cdot \mu m)$; $L_{max\lambda}$ – amount of radiation incoming to sensor, which ,after scaling, is converted into Q_{calmax} , $W/(m^2 \cdot ster \cdot \mu m)$; Q_{calmin} – minimal calibrated value of DN (0 or

1); Q_{calmax} – maximal calibrated value of DN (e.g., 255); Q_{cal} – calibrated value of DN.

Landsat Level 1 data are generated by various systems, thus influencing the value of parameter Q_{calmin}. Information on processing system is given in satellite image metadata file, values of the above parameters being given there also. For example, for Red, NIR and TIR- channels of sensors Landsat 5 and 7, values from Table 2 are used.

Table 2. $L_{min\lambda}$ and $L_{max\lambda}$ values for sensors Landsat 5 and 7.

Scanner	Band	$L_{min\lambda}$	$L_{max\lambda}$	Spatial resolution, m
Landsat-7 ETM+	Red, 3 (Low gain)	-5.0	234.4	30
	Red, 3 (High gain)	-5.0	152.9	
	NIR, 4 (Low gain)	-5.1	241.1	
	NIR, 4 (High gain)	-5.1	157.4	
	TIR, 6.1 (Low gain)	0.0	17.04	60
	TIR, 6.2 (High gain)	3.2	12.65	
Landsat-5 TM	Red, 3	-1.17	264.0	30
	NIR, 4	-1.51	221.0	
	TIR, 6	1.2378	15.3030	120

If metadata file is missing, then special calibration parameter files (CPF) are used, which are available in server Landsat: http://edclpdsftp.cr.usgs.gov/pub/data/CPF/. It should be noted, that specific period of time is ever associated with a specific sets of parameters and CPF. As it is known, imaging from Landsat has two options – Low or High. Metadata file also provides information on which of the gains and for what sensor the option was engaged.

After values have been obtained expressed in spectral radiance units, they are revaluated into brightness at-sensor temperature values (3). Atmospheric correction is often accounted for.

$$T_b = \frac{K_2}{\ln\left(\frac{K_1}{L_\lambda}+1\right)}, \qquad (3)$$

where T_b – brightness at-sensor temperature, K; K_1 и K_2 – calibration invariable of sensors Landsat (Landsat 2008). For Landsat-5 TM: K_1 = 607.76; K_2 = 1260.56. For Landsat-7 ETM+: K_1 = 666.09; K_2 = 1282.71.

Data sourced from sensor ETM+ before December 20, 2000, feature the error explained by wrong calibration of channel 6, thus all the values are 3 degree K greater than actual ones (Landsat 2008). To compensate for this inaccuracy, 0.31 value is subtracted from L_λ

obtained. This is a prerequisite for data processed at Level 1R and 1G

If atmospheric correction has not been done, the temperature is calculated by the following equation (sourced from: The Yale Center for Earth Observation):

$$T_b = \frac{K_2}{\ln\left(\frac{K_1 \cdot \varepsilon}{L_\lambda}+1\right)}, \qquad (4)$$

where ε – emissivity, a spectral radiation factor.

In course of quantitative analysis of temperature values calculated from data Landsat-5 TM in environment ENVI (with built-in computation algorithms), and with raster calculator from GIS GRASS, it has been established that optimal value is $\varepsilon = 0.956653$. Further we will demonstrate that it is with this factor value that data T_b, calculated both inside and outside environment ENVI, possess minimal values.

In estimation of reflectance most commonly employed method is TOA (top-of-atmosphere), which is based on expression (Chander, Markham & Barsi 2009):

$$\rho_\lambda = \frac{\pi \cdot L_\lambda \cdot d^2}{ESUN_\lambda \cdot \cos\Theta_s}, \qquad (5)$$

where ρ_λ – global reflectance TOA (nondimensional value); d – the Earth to the Sun distance (in astronomical units); $ESUN_\lambda$ – average extraterrestrial radiation – solar constant for each channel (amount of energy in its spectral band, $W/(m^2 \cdot ster \cdot \mu m)$); Θ_S – zenith angle of the Sun with respect to mapped area center in the moment of imaging (in degrees).

Table 3. Value of $ESUN_\lambda$.

Scanner	Value of $ESUN_\lambda$ for the band scanner, $W/(m^2 \cdot ster \cdot \mu m)$							
	1	2	3	4	5	6	7	PAN
Landsat-5 TM	1983	1796	1536	1031	220	–	83.44	–
Landsat-7 ETM+	1997	1812	1533	1039	230.8	–	84.90	1362

The authors have employed 2 methods for evaluation of LST.

1. Method suggested in works (Artis & Carnahan 1982; Nichol 1994, 2005; Xiong et al. 2012; Weng 2001; Weng, Lu & Schubring 2004; Wan 1994).

2. Methods based on algorithm mono-window (Qin, Karnieli & Berliner 2001).

The following expression underlays the first method of LST values estimation:

$$T_s = \frac{T_b}{1 + \left(\lambda \cdot \dfrac{T_b}{\rho}\right) \cdot \ln \varepsilon}, \qquad (6)$$

where T_s – surface level temperature value, K; T_b – at-sensor temperature calculated by (3) or (4), K; λ – average wave length of TIR-channel taken for calculations, μm; $\rho = 1.438 \times 10^{-2}$ м*K; ε – surface spectral radiation factor, calculated from NDVI value obtained with methods (Lim et al. 2012; Nichol 2005; Sun, Tan & Xu 2010; Li et al. 2004).

In the case of the algorithm mono-window for the 6-th band scanner Landsat-5 TM use the expressions (Qin, Karnieli & Berliner 2001; Liu & Zhang 2011):

$$T_s = \frac{a_6(1 - C_6 - D_6) + T_b(b_6(1 - C_6 - D_6) + C_6 + D_6) - D_6 T_a}{C_6}, \quad (7)$$

$$C_6 = \varepsilon_6 \tau_6, \quad D_6 = (1 - \tau_6) \cdot [1 + (1 - \varepsilon_6) \cdot \tau_6],$$

$$T_a = 16.0110 + 0.92621 \cdot T_0, \qquad (8)$$

where a_6, b_6 – experimental evidence constants: $a_6 = -67.355351$, $b_6 = 0.458606$; ε – spectral emissivity, calculated from NDVI

value; τ_6 – atmospheric spectral transmission factor, established with account for water vapor; T_a – effective average temperature of atmosphere; T_0 – near-surface temperature of atmosphere measured at 2 meters height (weather station). Determination of coefficients for (8) and calculation of τ_6 are described in detail in the works (Qin, Karnieli & Berliner 2001; Liu & Zhang 2011).

Expressions (1) – (8) also hold for data of sensors Landsat TM/ETM+. In some instances, like for sensor Terra/Aqua MODIS, ultimate products are used.

In estimation of surface temperature distribution the following Terra MODIS products are of interest: MOD11A1 (daily LST), MOD11A2 (eight-day LST), MOD11C3 (global LST monthly map), as well as MOD11_L2 (LST per 5-minute imaging). The last product mentioned has been preferred for suggested exploration of surface temperature pattern in urban territory.

Geometric correction of MODIS data is carried out with special utility programs: for processing levels 2G, 3 and 4 – MODIS Reprojection Tool (MRT); for levels 1B and 2 (MOD11_L2) – MODIS Reprojection Tool Swath (MRT Swath). Both are Java applications and require Java Virtual Machine.

To obtain ultimate temperature values in ^0C, in all above mentioned MODIS products the following expression is used:

$$T_{i,j} = (V_{i,j} \cdot 0.02) - 273.15, \qquad (9)$$

Values of d can be outsourced from (Chander, Markham & Helder 2009).

Value of $ESUN_\lambda$ for sensors TM/ETM+ are given in Table 3.

Value of zenith angle $\Theta_S = 90,0 - SE$, where SE is the Sun elevation angle value as of the moment of imaging (from metadata file).

where $T_{i,j}$ – temperature of (i,j)-th pixel, ^{0}C; $V_{i,j}$ – value of (i,j)-th pixel in the image containing LST data from MODIS sensor.

It should be pointed out that MOD11C3 product has low spatial resolution (5600 m), since it is meant for global territory coverage. Others feature spatial resolution of 1000 m. That is far from sufficient to construct fine UHI maps of smaller size territories, however, they are excellent tools for global monitoring of vast surfaces.

For finite LST maps charting it is convenient to use, alongside with MRT utility programs, software tools for operations with rasters, such as: gdal_calc.py out of library package GDAL, raster calculator Quantum GIS, module r.mapcalc in GIS GRASS, soft tool Model Maker in ERDAS Imagine and Band Math in ENVI etc.

4 RESULTS

Since data MOD11_L2 after processing with MRT Swath are presented by geographical projection on datum WGS-84 (code EPSG:4326), then all 5 satellite images of Landsat-5 TM are brought to this projection.

In course of investigation and with application of expressions (2) – (8), 5 LST maps have been obtained based on Landsat-5 TM data, as well as data of Terra MODIS MOD11_L2 have been used to construct LST maps with involvement of expression (9).

In Fig. 2 a scatter plot is presented to demonstrate difference between LST temperatures obtained with MOD11_L2 data and those drawn out of equation (4).

For assessment of LST determination results yielded by both ways, 10 control points were selected within target territory (Fig. 3). Results of temperature evaluation by different techniques appear in Table 4, where: Tb – temperature found by expression (3), Ttm – temperature found by expression (4), $Tmodis$ – temperature found by (9), $Tlst1$ – temperature found by expression (6), $Tlst2$ – temperature evaluated by (7), $Tenvi$ – temperature calculated with tools of ENVI environment. Fig. 4 represents differences in estimated temperatures for satellite image dated 22.04.2011.

On the assumption of temperature estimation results, obtained by different methods it has been verified, that the least differing were values of Ttm and $Tenvi$. Values of $Tmodis$ display comprehensive differences from calculated temperatures.

On the basis of sensors MODIS and TM data, linear regression factors and mean root square errors were found (Tables 5 and 6). Linear regression equation:

$$\hat{T}_{tm} = b_0 + b_1 \cdot T_{modis} , \qquad (10)$$

$$\hat{T}_{modis} = b_0 + b_1 \cdot T_{tm} , \qquad (11)$$

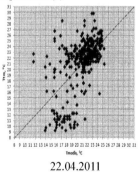

22.04.2011

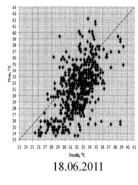

18.06.2011

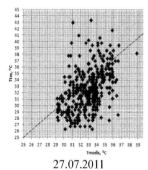

27.07.2011

155

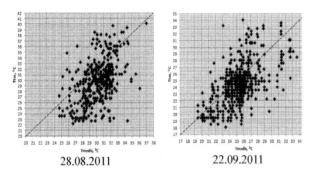

28.08.2011   22.09.2011

Figure 2. Scatter diagram of temperature values estimated on the basis of Terra MODIS data (axis X) and Landsat-5 TM data (axis Y).

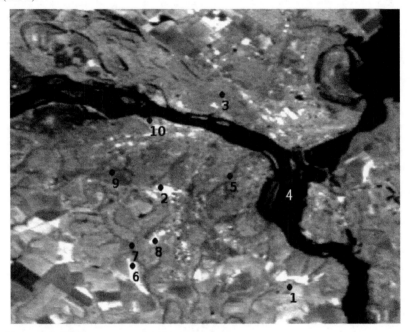

Figure 3. The control points temperature measurement.

Table 4. Temperature values estimated by 10 control points.

Point number	Tb, ^0C	Ttm, ^0C	Tmodis, ^0C	Tlst1, ^0C	Tlst2, ^0C	Tenvi, ^0C	ε
1	23.6826	26.7665	24.4500	24.6599	31.2848	26.3243	0.986212
2	30.4194	33.6409	21.5100	32.6711	41.0080	33.1542	0.97
3	20.1746	23.1879	22.5699	21.1426	26.2217	22.7700	0.986016
4	8.9817	11.7745	16.1499	10.9255	10.0671	11.4402	0.97
5	16.1028	19.0351	22.6299	17.0310	20.3449	18.6465	0.986208
6	26.6734	29.8181	23.6900	28.8696	35.6014	29.3558	0.97
7	17.0198	19.9702	22.1499	17.8913	21.6684	19.5749	0.987123
8	32.4556	35.7191	22.5300	34.7376	43.9468	35.2195	0.97
9	15.1786	18.0925	21.2299	16.1040	19.0110	17.7108	0.986162
10	51.3545	55.0196	14.1700	52.4676	71.2235	54.4089	0.986857

156

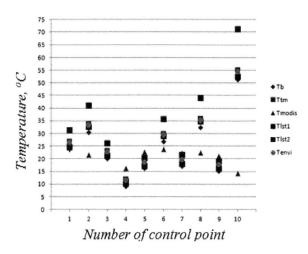

Figure 4. Scatter plot of temperature (axis Y), which are defined by different methods by ten points (axis X).

Table 5. Linear regression equation factors established on the basis of survey with application of (10).

Num.	Date of surveying	b_0	b_1	S_{err}*,^0C
1	22.04.2011	1.204194	0.953363	3.676092
2	18.06.2011	0.722270	0.944303	3.151843
3	27.07.2011	6.220470	0.796116	2.840840
4	28.08.2011	2.460166	0.905461	3.292144
5	22.09.2011	5.525097	0.747127	2.475026

* – mean root square error of estimation.

Table 6. Linear regression equation factors established on the basis of survey with application of (11).

Num.	Date of surveying	b_0	b_1	S_{err},^0C
1	22.04.2011	13.677127	0.340517	2.196984
2	18.06.2011	21.067553	0.356106	1.935523
3	27.07.2011	24.595944	0.258908	1.620060
4	28.08.2011	22.438855	0.260656	1.766352
5	22.09.2011	13.388179	0.486613	1.997444

By using linear regression equation (10), we have revaluated values of LST temperatures of Landsat-5 TM data to those of MOD11_L2. By using (11) back-calculation has been done. Mean root square error of estimation is characteristic of deviation of actual data from regression line. That makes possible the estimation of observation points with respect to regression right line.

Thus, if you want to adjust the values of the MODIS on the values of TM scanner, then it is advisable to use regression.

In course of investigation based on 5 satellite images of Landsat-5 TM a finite LST map of the field plot has been constructed (Fig. 5).

157

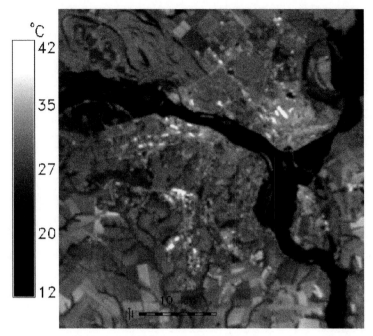

Figure 5. Finite map of thermal field of the city of Dnipropetrovs'k (data of sensor Landsat-5 TM).

5 CONCLUSIONS

In the research with application of GIS-tools of library GDAL, GIS GRASS and ENVI, the computation algorithms have been realized for construction of digital LST maps for the city of Dnipropetrovs'k, period 22.04.2011 – 22.09.2011. The comparison undertaken of computed results revealed similarity of temperatures *Tb*, *Ttm*, *Tlst1*, *Tenvi*, most comprehensive similarity being between *Ttm* and *Tenvi*. The difference between result sets of these temperatures averaged around 4 ^{0}C. In 10 control points selected temperatures *Ttm* and *Tenvi* differed by average 0.44 ^{0}C. Difference between *Tmodis* and *Tlst1* was 4.5 ^{0}C on the average.

Distinctions of temperatures *Tlst2*, established by mono-window method, are explained by a great number of additional parameters and complexity of atmospheric transmission spectral factor calculation, namely, with method of identification of water vapor in atmosphere and computation of average atmosphere temperature. In 10 points temperatures *Tlst2* exceeded *Tlst1* by 6.4 ^{0}C on the average.

However, construction of LST map provides accurate enough identification of urban construction zones which display quite high thermal background. Areas with the least LST values correspond to parks or ravine territories.

REFERENCES

Walawender, J. & Haito, M. 2009. *Assessment of thermal conditions in urban areas with use of different satellite data and gis.* EUMETSAT meteorological satellite conference: Bath: UK: CD-ROM.

Oguro, Y., Ito, S. & Tsuchiya, K. 2011. *Comparisons of brightness temperatures of Landsat-7/ETM+ and Terra/MODIS around Hotien Oasis in the Taklimakan Desert.* Applied and Environmental Soil Science, Vol. 2011, article ID 948135: 11.

Qin, Z., Karnieli, A. & Berliner, P. 2001. *A mono-window algorithm for retrieving land surface temperature from Landsat TM data and its application to the Israel-Egypt border region.* International Journal of Remote Sensing, Vol. 22, Issue 18: 3719-3746.

Nichol J. 2005. *Remote sensing of Urban Heat Islands by day and night.* Photogrammetric Engineering & Remote Sensing, Vol. 72, Issue 5: 613-621.

Barsi, J.A., et al. 2003. *Landsat TM and ETM+ thermal band calibration.* Canadian Journal of Remote Sensing, Vol. 29, Issue 2: 141-153.

Artis, D.A. & Carnahan, W.H. 1982. *Survey of emissivity variability in thermography of urban areas.* Remote Sensing of the Environment, Vol. 12: 313-329.

Xiong, Y., Huang, S., Chen, F., Ye, Hong, Wang, C. & Zhu, C. 2012. *The impacts of rapid urbanization on the thermal environment: a remote sensing study of Guangzhou, South China.* Remote sensing, Vol. 4: 2033-2056.

Nichol, J.E. 1994. *A GIS based approach to microclimate monitoring in Singapore's high rise housing estates.* Photogrammetric Engineering & Remote Sensing, Issue 60(10): 1225-1232.

Weng, Q., Lu, D. & Schubring, J. 2004. *Estimation of land surface temperature-vegetation abundance relationship for urban heat island studies.* Remote Sensing of Environment, Vol. 89: 467-483.

Weng, Q., 2001. *A remote sensing-GIS evaluation of urban expansion and its impact on surface temperature in the Zhujiang Delta, China.* International Journal of Remote Sensing, Vol. 22(10): 1999-2014.

Wan, Z., 1994. *MODIS land surface temperature algorithm theoretical basis document (LST ATDB).* Version 1, Contract No. NASS-31370: 37.

Landsat Project Science Office. 2008. *Landsat-7 science data user's handbook.* Goddard Space Flight Center, NASA: Greenbelt, MD, USA: 186

Chander, G., Markham, B.L. & Barsi, J.A. 2009. *Revised Landsat-5 thematic mapper radiometric calibration.* IEEE Geoscience and Remote Sensing Letters, Vol. 4(3): 490-494.

Chander, G., Markham, B.L. & Helder, D.L. 2009. *Summary of current radiometric calibration coefficients for Landsat MSS, TM, ETM+, and EO-1 ALI sensors.* Remote Sensing of Environment, Vol. 113: 893-903.

Lim, H.S., Mat Jafri, M.Z., Abdullah & Sultan Alsultan, K. 2012. *Application of a simple mono window land surface temperature algorithm from Landsat ETM+ over Al Qassim, Saudi Arabia.* Sains Malaysiana, Vol. 41(7): 841-846.

Sun, Q., Tan, J. & Xu, Y. 2010. *An ERDAS image processing method for retrieving LST and describing urban heat evolution: a case study in the Pearl River delta region in South China.* Environ Earth Sci, Vol. 59: 1047-1055.

Li, Fuqin et al. 2004. *Deriving land surface temperature from Landsat 5 and 7 during SMEX02/SMACEX.* Remote Sensing of Environment, Vol. 92, Issue 4: 521-534.

Liu, L. & Zhang, Y. 2011. *Urban Heart Island analysis using the Landsat TM data and ASTER data: a case study in Hong Kong.* Remote Sensing, Vol. 3: 1535-1552.

Energy Efficiency Improvement of Geotechnical Systems – Pivnyak, Beshta & Alekseyev (eds)
© 2013 Taylor & Francis Group, London, ISBN 978-1-138-00126-8

Predicting methane accumulation in the Donetsk coal basin (Ukraine) on the basis of geological, geophysical and space data

B. Busygin & S. Nikulin

State Higher Educational Establishment «National Mining University», Dnipropetrovs'k, Ukraine

ABSTRACT: The paper presents computer technology of forecasting methane accumulations from coal deposits based on integrated analysis of ground-based, borehole and space data. The results of forecasting of the perspective zones in the Donetsk Coal Basin (Ukraine) deposit are described.

1 INTRODUCTION

Production of non-attached methane from coal beds is a perspective trend for getting ecologically pure and effective power resources, and at the same time the solution of the important problem of mines safety.

By various estimations, the resource of coal methane amounts to 8-13 bln. cubic meters in the Donetsk coal basin. Up to 4 billion cubic meters of methane are exuded from bowels annually, but extraction does not exceed 8 % of this quantity. Therefore the problem of forecasting mine fields methane accumulation is of primary importance.

Traditional methods of new methane accumulations forecasting are based on the direct study of material and properties of rocks. They require large mining efforts and are time and fund consuming. Therefore great attention is paid to creation of methods and technologies that use relatively inexpensive ground-based geophysical techniques and satellite imagery allowing to predict the deposits of methane.

This paper describes the geoinformation technology which allows to forecast the sites with anomalous accumulations of methane in the Donetsk Coal Basin (Donbass) mine fields using space data and ground-based observations. The technology provides joint analysis of complex heterogeneous and different-level geological, geophysical and satellite data. It is based on the methods of Data Mining, image processing and lineament analysis (Pivnyak 2007; Busygin 2009).

2 DESCRIPTION OF THE INVESTIGATED AREA

The size of the investigated area is 9.1×7.0 km. It covers the mine field of Zasyadko operating mine (Fig.1). The investigated area is a hilly plain with a height difference ranging from 155 to 272 m above sea level. The landscape has been changed due to intensive anthropogenic factors; about 10% of the territory is occupied by mining facilities.

The geological structure of the area is formed by deposits of the Middle Carboniferous C_2^5, C_2^6 and C_2^7 suites. They are represented by alternating sandstone, siltstone and mudstone layers of different thickness containing thin layers of limestone and coal.

There are a lot of folds and faults of submeridional and sublatitudinal direction in the central and eastern parts of the area. The investigated area is characterized by a complex structure and a large number of tectonic faults with amplitudes of 20-50 m. The presence of low-amplitude tectonics complicates greatly the structure of the coal beds. Numerous zones of metamorphized rocks and fracture zones are sensitive to infiltration of hydrothermal solutions, natural watering and are accompanied by anomalies of radioactive and methane - carbonic acid gases.

Figure 1. Location of Zasyadko mine

The strike of the coal-bearing rock mass is close to sublatitudinal, with the north-western angle of dip of 3-18°. Dip angle of rocks reaches 60° in the area of flexural folds. The m_3 and l_4 beds located at depths of up to 1200 m are currently mined. Coals of these beds are inclined to spontaneous ignition. Free methane is concentrated in local folds of sandstone beds overlaid by poorly permeable limestones and mudstones. Known concentrations of methane are confined to the faults registered by ground-based geophysical methods. The $m_4{}^0Sm_4{}^1$ sandstone bed located between the m_3 coal bed and the M_5 limestone bed is the most favorable.

3 INITIAL DATA

During the prediction process in the investigated mine field we have used:
- geological maps and schemes with the scale 1:25 000;
- areal surveys of gravitational field with the scale 1:25 000;
- digital elevation model (DEM) constructed using SRTM radar space imagery (Werner 2001);
- gas content data of 203 boreholes located within the mine field;
- maps of the $m_4{}^0S_4{}^1$ sandstone properties created on the basis of boreholes measurements: bed thickness, porosity factor etc.
- information about coal beds and limestone overlying beds.

Prior studies (Goncharenko 2007) indicated the following favorable factors in searching for methane potential accumulations:
- local elevations on the m_3 bed horizon;
- higher values of relative thickness factor of the $m_4{}^0S_4{}^1$ sandstone;
- higher values of thickness of the $m_4{}^0S_4{}^1$ sandstone;
- higher values of porosity factor of the $m_4{}^0S_4{}^1$ sandstone;
- local minima of gravitational field.

4 TECHNOLOGY STEPS

The technology is based on the RAPID GIS - a specialized geoinformation system (Pivnyak 2007) that enables prediction of various objects and phenomena by a set of direct and indirect features using processing and integrated analyzes of different nature data (geological, geophysical, geochemical, landscape), obtained from various sources (boreholes, ground, space).

Data Mining methods play a key role in prediction tasks. Prediction technology in the RAPID GIS is based on successive stages: data pre-processing, training and control sets forming, feature space creation and optimization (Busygin 1991), supervised classification and prediction quality assessment.

Data pre-processing. All geological, tectonic and structural maps are being georeferenced and digitized. Map objects (such as faults, geological boundaries, hole mouths etc.) are grouped into

162

individual thematic vector GIS-layers and then converted into a grid form by rasterization.

Grid data (geophysical fields, satellite images ets.) are checked for errors and united into integrated grid network. Satellite images are subjected to geometric correction and various transformations, such as histogram equalization, brightness and contrast change etc.

On geophysical maps, geological lineaments (linear features) of digital elevation models and satellite images are distinguished. Most of them correspond to different tectonic breaches of the Earth's crust. Lineament study is of much value for the new accumulations prediction not only in the Donbass, but also other coal basins (Ayers 1994; Mazumder 2004; Solano-Acosta 2007). Lineaments are detected automatically on the basis of Kenny optimal detector (Canny 1986) and Hough transformation (Sewisy 2002), after which they are examined and refined by an expert geologist in an interactive mode.

Training set formation. The suggested technology widely uses methods of Data Mining; primarily supervised machine learning, which involves using the training sets. The grid cells located above the wells mouths are used as training set objects (samples).

Thus, 149 boreholes with depth up to 1500 meters were used on the territory of the investigated mine field. Commercial gas accumulations have been detected in 9 of them. Grid cells located above these boreholes formed the first training set. The others composed the second set (140 samples).

Then, the samples of both sets were analyzed using special methods to assess the compactness of their location in the multidimensional feature space (Busygin 1991) and, if necessary, were transferred into another set or into test set used to assess the strength and quality of prediction results.

Transformations calculation and feature space formation. Feature space consists of a set of grid map layers describing the territory and various aspects of gas content and related geological structures. As shown by the corresponding study, the initial features (gravitational field, DEM, coal bed characteristics and sandstone bed characteristics) are not sufficiently informative for the reliable forecast. For more features, it is necessary to calculate their various transformations. The RAPID GIS enables to perform more than 200 transformations, including differential, texture, histogram, morphologic, correlation, fractal, etc. The evaluated features of lineaments network reflect structural and tectonic situation of the area (Pivnyak 2010).

Currently, there is no formal procedure for optimal informative feature system determining. Therefore, at first the initial feature system is being created. It consists of a set of measured satellite, geological and geophysical features as well as their transformations. Subsequently the resulting feature space is minimized by discarding the irrelevant features. Minimization is performed in two stages. During the first stage the uninformative or redundant features are discarded on the basis of visual analysis. During the second stage, the informative feature system is created, and the decisive rule of classification is formulated. Such rule minimizes a certain classification quality functional (commonly – an error of classification). During this stage the previously formed training sets are used. Features that were not included in the informative set are not used in further calculations.

Prediction and quality assessment. Previously determined decisive rule and informative feature system allow to perform classification including recognition and ranking. Recognition assigns each point of the territory to one of the classes which correspond to training sets. Ranking assesses the measure of similarity between the territory and one of the training sets.

The results are represented in the form of predictive maps and charts that reflect promising areas in terms of methane abnormal concentrations in the rock mass.

The necessary step is quality assessment of the results obtained. The objects of the test set are used for this purpose. They correspond to boreholes with commercial gas content which were not used in training sets. To assess the validity of forecasting results a variety of quality indicators have been calculated. In case of unsatisfactory assessments, it is recommended to conduct additional studies of training sets and change the feature system.

5 METHANE ACCUMULATIONS FORECASTING WITHIN ZASYADKO MINE FIELD

The described technology was used to predict the new accumulations of methane.

The input features used for classification were:

F1 – gravitational field Vz;

F2 – amplitude of m_3 coal bed local structures;

F3 – $m_4^0 S_4^1$ sandstone thickness;

F4 – $m_4^0 S_4^1$ sandstone porosity factor;

F5 – m $_4^0S_4^1$ sandstone relative thickness factor;

F6 – sandstone total thickness within the interval;

F7– surface heights, derived from radar imagery.

The features were adapted to 20x20 m grid and normalized by [0..1] range.

Many transformations were calculated for geophysical data (Busygin 1991). Many of them are more informative than the original features. Spatial features of the area lineament network were calculated separately. For this, lineaments of the gravitational field (the scale of 1:25 000 and 1:10 000) and DEM were distinguished in interactive mode. It should be noted that the lineaments of DEM and gravitational fields are closely related, the overlapping and forming a network with 4 predominant directions – 0, 45, 90 and 135° (Fig. 2). The DEM lineaments reflect, first of all, heterogeneity of the surface layers, whereas deeper objects are reflected in the gravitational field. The combination of both types of lineaments provides a fairly complete picture of the fracture field of the studied area. The results are shown in Figure 2d.

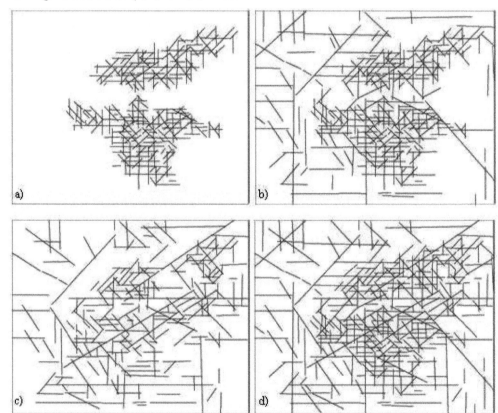

Figure 2. Lineaments allocated by: a) gravitational survey in 1:10 000 scale and b) the gravitational surveys in 1:10 000 and 1: 25 000 scales, c) digital elevation model, and d) gravity surveys and digital elevation model

After constructing the lineaments network, we could calculate their different characteristics in the moving window size of 140x140, 300x300 and 400x400 m, e.g. the total length of lineaments per area unit (density), the predominant azimuth, the ratio of total length and number of lineaments etc. In addition, these characteristics were computed for gravitational field and topography lineaments combined into a single set.

During the next step of research, the training set of "promising" objects was analyzed. The method of multidimensional scaling, which allows to display multidimensional information in three-dimensional space (Borg 2005; Busygin 2009) proved that the samples of the training set fall into several isolated groups (Fig. 3). This is explained by the fact that the boreholes are located in different geological conditions, and moreover, accumulations of methane are associated with different structural formations.

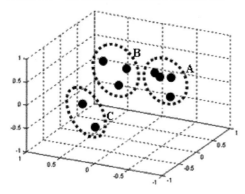

Figure 3. Dividing the objects of the training set into A, B, and C subsets in a multidimensional feature space, reduced to a 3-dimensional one

It was decided to divide the "promising" training class into several subsets. For this, clustering of samples within previously formed feature space was made. It was carried out by various methods - k-means, IZODATA, Kohonen neural network. All calculations were performed in two ways – in the first approach we clustered only boreholes with proved gas content, in the second we considered all 149 boreholes. The complex analysis of the results showed that the best way is to divide 9 "promising" samples into three subsets - A, comprising 4 samples, B (3 samples), and C (2 samples). The samples representing the boreholes without gas formed the D set ("empty") (Fig. 4).

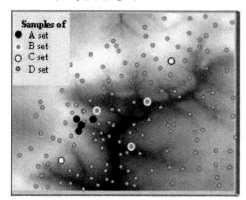

Samples of
● A set
• B set
○ C set
◎ D set

Figure 4. Location of training set samples (superimposed on a digital elevation model)

Next, using special methods of informative feature systems construction, we created 3 sets for A, B and C classes. Features included in the classes, provide the best criteria for separating the "promising" training set samples from the "empty" D set.

The informative feature system for A set includes:

1. $m_4{}^0S_4{}^1$ sandstone relative thickness factor (F5).

2. Kendall's rank correlation coefficient between the gravitational field and the $m_4{}^0S_4{}^1$ sandstone thickness.

3. The sum of the absolute values | Aij | from the upper triangle of the matrix formed by signs pairs covariances F1..F7, j <i.

4. Characteristic, reflecting the parameter of the gravitational field histogram (Busygin 1991).

5. The direction of the horizontal gradient of the $m_4{}^0S_4{}^1$ sandstone porosity factor calculated in a moving window 160x160m.

The error of classification for the set is 5.56%.

The informative feature system for B set includes:

1. Surface heights, derived from radar imagery (F7).

2. Integral characteristic of DEM.

3. Kendall's rank correlation coefficient between the gravitational field and the $m_4{}^0S_4{}^1$ sandstone thickness.

4. The characteristic reflecting parameter of the gravitational field histogram.

5. The total length of the lineaments of the gravitational field and topography in a moving window 300x300m.

The error of classification for the set is 9.35%.

The informative feature system for C set includes:

1. $m_4{}^0S_4{}^1$ sandstone thickness F3.

2. Textural characteristics of the gravitational field, which reflects the inverse moment of difference (Haralick 1979) calculated in a sliding window 460x460m.

3. The characteristic reflecting parameter of the gravitational field histogram.

4. The direction of the horizontal gradient amplitude of the coal bed M_3 local structures in a moving window 140x140m.

5. The direction of the horizontal gradient of the $m_4{}^0S_4{}^1$ sandstone porosity factor calculated in a moving window 160x160m.

6. The total length of the lineaments of the gravitational field and topography in a moving window 400x400m.

7. The length of the lineaments of the gravitational field in a sliding window 300x300m.

The error of classification for the set is 6.72%.

As a result of feature sets used for classification, the maps showing the similarity of territory to samples with A, B and C classes were created. The maps were combined by calculating the weighted average of the T in the regular grid nodes using:

$$T_{ij} = 0.45*A_{ij} + 0.33*B_{ij} + 0.22*C_{ij}$$

where i, j – the indexes of the grid nodes, A_{ij}, B_{ij}, C_{ij} - the corresponding values of a similarity measure with the classes A, B and C. The values of weights were assumedproportional to the number of samples in a set.

The map created reflects integrated prospectiveness of the territory for methane accumulations detection (Fig. 5).

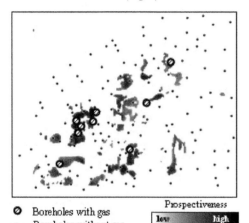

○ Boreholes with gas
· Boreholes without gas

Prospectiveness
low   high

Figure 5. Map of prospectiveness of new methane accumulations detecting

The reliability of the constructed maps is proved by the fact that all objects of classes A, B, C locate within the zones with the highest values of similarity, while the overwhelming majority of objects of class D are beyond their borders.

6 CONCLUSIONS

- The use of the proposed technology in the Central Donbass region yielded forecast maps to identify the areas prospective for discovering methane accumulations. The obtained results agree well with the geological facts and verified by drilling.
- Of particular interest are the results of lineament analysis. As seen in Figure 6, the majority (8 of 9) of boreholes with the confirmed

gas content are confined to areas of high concentration of lineaments and thus to fractured rocks. This must occur due to the fact that these deconsolidated zones have a higher collector properties and function as gas migration channels. In addition, the relief of the current tectonic stresses occurring along the ancient lineament network increases rock fracture and releases "tied" gas.

Hence, high concentration of lineaments can be considered an important feature for predicting coal basins methane accumulations. This, first of all, relates to the lineaments of the gravitational field which is not exposed to man-made changes and better reflects deeper objects and processes in comparison with DEM and satellite imagery.

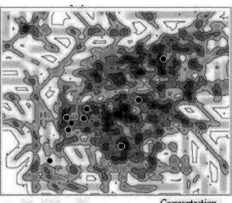

· Boreholes with gas
· Boreholes without gas

Concentration
low   high

Figure 6. Map of total length of lineaments per unit area (concentration) calculated with sliding window 300x300m

- Research has demonstrated high informative value of the data obtained from relatively inexpensive geophysical surveys and radar satellite imagery. Thus, 11 of the 17 features included in the informative systems for A, B, C classes are derivatives of the gravitational field or the relief.

Computational experiments showed that results of gravity surveys utilizing in forecasting process in addition to geological data, allows to reduce the error of classification by 30-50%. If the satellite and the relief data are also used, the error is reduced by 5-15% additionally.

To conclude, the obtained results indicate the importance and feasibility of integrating geological, geophysical and satellite surveys data to detect accumulations of methane in the Central Donbass. Such integration allows to greatly improve the reliability of forecast without significant expenditure growth.

REFERENCES

Borg, I. & Groenen, P. 2005. *Modern multidimensional scaling: theory and applications (2nd ed.).*Springer-Verlag: 614.

Busygin B., Miroshnichenko L. 1991. *Pattern recognition for geological and geophysical prognosis.* Dnipropetrovs'k: DMI: 121.

Busygin B., Nikulin S. 2009. *The methodology of oil and gas deposits prognosis by space and geophysical data.* Proceedings of 71rd EAGE conference : Amsterdam: The Netherlands: 4.

Canny J. F. 1986. *A computational approach to edge detection.* IEEE Trans Pattern Analysis and Machine Intelligence, Vol.8: 679-698.

Goncharenko V., Svistun V., Gerasimenko T., Malinovskiy A. 2007. *The basis of complex prognosis of methane accumulation zones in Donbass using geological and geophysical data.* Geologist of Ukraine, Vol.1: 36-43.

Haralick R.M. 1979. *Statistical and structural approaches to texture.* Proceedings of the IEEE, Vol.67: 786-804.

Pivnyak G., Busygin B., Nikulin S. 2010. *Geoinformation system RAPID as the means of solving the problems of environment and nature management.* Proceedings of SWEMP-2010: Prague: Czech Republic.

Pivnyak, G. G., Busygin, B. S., Nikulin, S. L. 2007. *GIS-technology of integrated heterogeneous and different-level geodata analysis.* Reports of the National Academy of Sciences of Ukraine, Vol. 7: 121-128.

Sewisy A.A. 2002. *Graphical techniques for detecting lines with the Hough transform.* Int. J. Comput. Math, Vol.79: 49-64.

Werner M. 2001. *Shuttle Radar Topography Mission (SRTM): mission overview.* Frequenz: J. Telecom, Vol.55: 75-79.

Energy Efficiency Improvement of Geotechnical Systems – Pivnyak, Beshta & Alekseyev (eds)
© 2013 Taylor & Francis Group, London, ISBN 978-1-138-00126-8

High-energy processing of materials-technologies of the 21-st century

R. Didyk
State Higher Educational Institution "National Mining University", Dnipropetrovs'k, Ukraine

ABSTRACT: The possibilities of effective use of high-module power sources including energy of explosion and powerful ultrasonic vibrators for manufacturing welding methods, hardening, synthesis of superstrength materials, and dimensional finishing of components are shown in the paper.

There are given data demonstrating science intensive production with the use of listed methods for the most important branches of industry.

1 INTRODUCTION

The use of high-module sources for manufacturing methods of metal working should be considered as one of the most significant achievements in the sphere of science and engineering. The peculiarities of working connected with the use of shock waves of different intensity permitted to develop absolutely new manufacturing methods of metal working, and improve the abilities of available methods.

Progress of methods of explosive working of metals advance a number of problems in the mechanics of rigid body being deformed, in materiology, and physics of the explosion. To use and analyze the phenomena of the explosion, and evaluate the behaviour of materials under the conditions of high gradients of pressure and load rates there were developed analytical and numerical methods of parameters of influence determination as well as modern ways of rapid processes. Fundamental works in this field give ability to develop a number of effective manufacturing methods for metal working.

Thus the energy of the explosion is effectively used for forming operations of metal working with the help of pressure, cutting, welding and metal strengthening. Any patterns are made on the metal, polymer processes are accelerated, synthesis of diamonds is performed, high density compacts of power materials are produced, the processes of defect formation in superstrength materials are stimulated, the values of residual stresses of complex mechanical constructions (Didyk 1999) are checked with the help of the explosion.

2. EXPLOSION WELDING

During the first stages of research the unique abilities of explosion welding were determined. They give possibility to establish practically reliable links of any metals and alloys independently of their physical and mechanical properties.

Specialists believe that during the near decade qualitative improvement of properties of materials can be obtained not only by production upgrading and know alloys working but the composite, in particular laminated materials manufacture. Those laminated materials have not only high level of toughness and corrosion resistance but a high level of resistance to cracking. That's why explosion welding is very important for different constructions with significant resource of reliability and durability. The connection of metals by means of explosion is performed under the conditions of high-speed collision of interacting metals without previous heating under high energetic indices of source and deformation conditions. The intensive plastic deformation of contact surfaces is the basis of metal bonding while welding by means of explosion. It is accompanied by formation of stable or unstable jet, and this process is accompanied by wave formation at the edge of welding. Deformation of contact surfaces and jet formation and as the result the properties of joint area for the given metal couple under fixed conditions are determined by means of following parameters: a rated speed of collision, a speed of contact point, an angle between cladding layer and basis while rolling (Fig.1a,b).

a   b

Figure. 1. The Dynamics of Explosion Welding Process

a) impulse X-ray pattern of explosion welding moment after 22 microseconds the explosion has been initiated.

2 – fixed plate, 3 – movable plate, 4 – nondetonated part of explosive agent charge, 5 – detonator (official photo by the Institute of Hydrodynamics named after M.A. Lavrent'ev CO RAS).

b) a sketch of profile of the pattern after "sudden stop" of explosion welding process.

1 – movable (made of aluminium); 2 – the main (made of steel) plates; 3 – back jet (aluminium); 4 – layer of matter; 5 – explosion image at the moment of "sudden stop" of process.

c) macrographical borders of welding of multilayer metal construction.

The progress of science and technology and explosion welding industrial mastering helped to create efficient method of multilayer constructions for pipes and tubes production.

The range of sizes of bimetal tubes and pipes produced by means of explosion welding, and sphere of their application are shown in the Table 1.

Table 1. Range of sizes of bimetal piper/tubes produced by means of explosion welding

#	Bimetal pipes/tubes	Field of application
1	st. 10+copper	Chemical industry, navy
2	X18H10T+AMГ-6 X18H10T+BT5	Rocket production
3	Copper+niobium+copper	Fiber optics
4	X18H10T+zirconium	Nuclear power engineering
5	АД1+copper	Power engineering
6	St. 20+X18H10T	Engineering
7	38ХН3МФ+ЭП131	Engineering
8	АД+X18H10T	Aviation
9	38ХН3МФА+ВХ-2К	Engineering

3. METAL REINFORCING BY MEANS OF EXPLOSION

The advantages of explosive deformation to compare with ordinary methods of deformation reinforcing shown as the lack of deformation or its residual presence, and isotrop reinforcement under great indices of durability have stimulated interest of specialist to physical metallurgy problems and tasks of explosion reinforcement.

The works which peculiarities were to study structural changes in the steels exposed to influence of different blasting materials played

important role. The results of the study show that structural changes stipulating high level of reinforcement under significant ductility of material can be produced if combined charges of blasting material with specific parameters are used.

As study shows the difference in the efficiency of metal reinforcement under contact explosion is connected with initial parameters of the border of "Blasting materials-metal" which are determined by the type of BM, terms of progress and interaction of detonation wave with obstacle, initial stale, and behaviour of material being

170

worked under condition of high dynamic pressure.

Tests concerning the influence of shock waves of cyclical strength of different steels showed high efficiency of explosion working. Fig.2.Demonstrates competitive data of various reinforcement workings of 40XH steel to the indices of cyclical strength. It is shown, that under chosen method of explosion reinforcement the durability limit of the given steel quality increases by 50 to 0 per cent.

There was developed and used the method of metal reinforcement by means of two-layer BM which layers differ greatly on their physical and chemical properties, an high efficiency of this method was demonstrated to compare with previous methods. There were determined criteria condition leaving out micro-and macrodistructions of material under its working with the help of explosion. There were studied detonation operation modes in combined charges with cylindrical and conic symmetry on the basis of which the range of optimum angles of detonation front contact with the maximal reinforcement effect is possible.

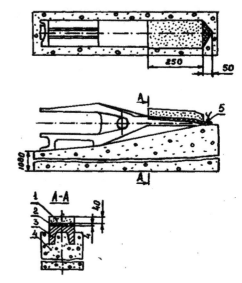

Figure.3. Flowsheet of explosion Reinforcement of ЭКГ-4.Excavator Dipper Teeth
1 – a charge of low-speed BM; 2 – a charge of rapid BM; 3 – an element being reinforced; 4 – a basis; 5 – a blasting cap

The research on the problems concerning metal reinforcement by means pf explosion the basis to develop the process of detail working which completely characterize heavy engineering industry of metallurgic and mining equipment (rolls, gearings of pull rope for walking excavator ШГ15/90, bucket teeth of excavator) were successfully used by the important enterprises of the country.

Fig.3. shows the procedure of explosion reinforcement of ЭКГ- 4 excavator bucket teeth.

The progress of activities aimed at steel explosion reinforcement is based upon the research evaluation of shock waves influence on residual stresses in the components having complex geometric shape.

The matter is that a number of factors connected with heterogeneity of product materials, and production process, and construction peculiarities stipulate sufficient reserve of residual stresses which greatly limit the service life of machine components.

The working by shock waves of given configuration and rate with the use of high-module BM which thickness is less then a millimetre improves greatly the state of stress. It is worth saying that tensile residual stresses decrease in the area of their concentration is up to 50 per cent and more of the initial value (Didyk, Loskytov & Aratskiy 2002).

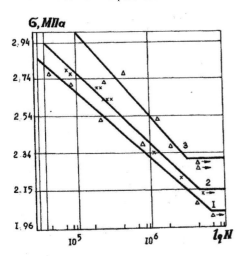

Figure.2. Competitive Evaluation of Steel 40XH Durability Limit under Various Ways of Reinforcement
1 – thermal working; 2 – nurling by rollers; 3 – explosive reinforcement

171

4. SYNTHESIS OF DIAMOND OF SUPERHARD MATERIALS REGENERATION IN SHOCK WAVES

The process of shock compression products preservation have been developed by the soviet and foreign researchers in the 1950s not only began a new stage in the physics of shock waves but also introduced wide-spread application of loading method itself. The production of diamonds and other superhard materials, the activation of powders by means of shock waves for further use in traditional methods are the most important achievements in this fields.

This trend is very interesting because it is stipulated by the necessity to develop theoretic ideas concerning the structure and properties of material under the influence of such natural extreme parameters as temperature and pressure as well as high energetic ways of influence application in modern technologies.

Fast increase and decrease of pressure, and significan concentration of energy stipulating structural and phase transformation during microsecond are physical properties of material shock compression.

The peculiarities of research in the University were determined by the development of different constructions of shock preservation and geometry of charge being used. It helps to perform plane, cylindric, conic or spheric types of loading of reactive powders and compact graphite containing materials. Morphologic peculiarities and physical properties of diamond crystals as well as the structures of graphitized materials were studied with the help of ЭПР method, and x-rays, and optic and electron microscopy. First there was studied the influence of various working methods (forging, explosion, thermocycling) of pig iron with different chemical composition and matrix structure on the parameters of carbon mass transfer.

The research concerning loading influence of solid alloy masses reactive ability showed that inner energy, and density of different defects, and their chemical activity increase greatly in the worked materials. The listed peculiarities let perform finely dispersed grinding of the alloy without physical and thermal, and metallurgical ways of regeneration. With it the period of powder grinding previously worked by explosion is more than 20times less to compare with current grinding method (Didyk 2004).

5. SURFACE ENGINEERING AND RENOVATION OF PRODUCTS

Surface modification to decrease greatly roughness and prolong the life of machine components is the result of high-energy dimensional ultrasonic component working with simultaneous introduction of geomodifiers of friction in surface layers of metal.

Geomodiers of friction (GMF) are the complex of grinded natural materials containing minerals of ultrabasic rocks being at the borders of tectonic platforms, and used by nature as the materials for triboprocesses. Specifically grinded and mechanically activated finely dispersed powders (0,5 to 1 mm) being in the friction area structurally modify working surface. These changes can modify it to become tribologically profitable, and their use helps to restore wornout working surfaces.

Shaft necks (with the diameter of 120mm) of 8HD pumps made of steel 30ХГСА were subjected to strengthening and finishing working at the integrated iron-and-steel works "Krivorozhstal". As compared with traditional methods of finishing operations (finishing turning and grinding) the high-energy dimensional ultrasonic working with simultaneous introduction of GMF in surface layers showed the improvement of durability five times as much and significant decrease in working surface roughness.

Measurements by roughness indicator "TAIYSURE-5" the microprofile of surface shown the following: the shaft sector roughness decreased as much as 3,5 times after it had been worked by ultrasonics, and when the surface was worked by ultrasonics with GMF together its roughness decreased as much as 6 times (Fig. 4).

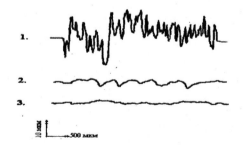

Figure.4. The Profile Diagrams of shaft surfaces (vertical enlargement is 1000, and horizontal one is 500)
1. initial one (after turning working); 2. after it had been worked by ultrasonics; 3. after joint working by ultrasonics and geomodifier of friction.

6 CONCLUSIONS

Given analysis illustrating the abilities of high-energetic metal working confirms that the main idea of the modern production processes is to provide the manufacture of machines and materials as the basis to solve the problem of quality and competitiveness of science intensive products in the world market.

REFERENCES

Didyk, R.P. 1999. *Theory and practice of handling explosive materials in the National Mining Academy of Ukraine* (in Russian). Dnipropetrovs'k: Scientific Bulletin of National Mining University, Issue 3: 56-60.

Didyk, R.P. 2004. *Regeneration of hard alloys in shock waves* (in Russian). Kharkiv: East European Journal of advanced technologies, Issue 3: 60-61.

Didyk, R.P., Loskytov, D.V. & Aratskiy P.B. 2002. *The method of hardening and finishing machining* (in Russian). Ukraine: Patent bul. №11.

Energy Efficiency Improvement of Geotechnical Systems – Pivnyak, Beshta & Alekseyev (eds)
© 2013 Taylor & Francis Group, London, ISBN 978-1-138-00126-8

Application of the optimal set partitioning method to problems of wireless network engineering

S. Us
State Higher Educational Institution "National Mining University", Dnipropetrovs'k, Ukraine

ABSTRACT: The optimal location problem for wireless network base stations in a given region has been studied. The aim of such location is to ensure the best signal quality. Discrete, continuous and combined mathematical models have been formed. A new solution algorithm, based on continual set partitioning method, has been proposed for a location problem with a discrete set of base locations and continual set of clients.

1 INTRODUCTION

One of the greatest achievements in the area of wireless networks is cellular communication. The role of this technology is just as great as the boom of personal computers in the 80's. A mobile phone has become a customary household item. Its prevalence exceeds the number of landline phones. Wireless networks today include hundreds of base stations. However, there are some problems related to the use of mobile communication, e.g. interruptions of communication, link failure, loss of words etc. (Mahmudov 2005; Vishnevskij 2003). These situations occur due to insufficient power of radio signal, which fades in the process of transmission from the base station to a mobile one. Thus, the question of wireless networks optimization remains important. Answering this question implicates a solution of various related problems. One from them is the optimal placement of base stations (BS) in a given territory. This problem can be formulated as the location-allocation problem.

The paper described mathematical models, methods, and the algorithm, that can be used to solve this problem.

2 BUILDING MATHEMATICAL MODELS AND ALGORITHMS

2.1 General assumptions

The problem of the optimal placement of base stations in a given area can be formulated as one of the research tasks.

Task 1. The number of base stations is known. They should be placed in the given territory so as to provide the best value of the quality criteria.

Task 2. The minimum number of base stations and their location in the given area should be determined, to ensure the predetermined value of the location quality criteria.

Task 3. There is a number of base stations in the given area, characteristics and location of which are known. One should determine the minimum number of additional base stations and their location in that territory to provide the required value for the location quality criteria.

The mathematical formulation of these tasks is optimal covering problem and different location problem. Different approaches to solving these problems were developed by (Piyavskij 1968; Galenko 2002; Baldacci et al 2002, 2008; Kiselova & Shor 2005; Kiselova et al 2009; Francis & Lowe 1992; Drezner & Hamacher 2001).

In this section we present some models appropriate to the optimal location problem for wireless network base stations in a given region. We make the following assumptions regarding the elements of the task to describe the model.

– All potential clients are connected to BS;
– Each client is connected to only one base station;
– Each base station can serve a predetermined number of clients only. It depends on the known capacity.

It is notable, that there are many quality indicators to characterize the work of the wireless network as a whole or its individual components. For example, they may be: the cost of the system (Galenko 2002), system capacity, requirements of compliance with environmental standards, and the signal level (Beresnev 2002). In this paper, we selected the loss of the signal as a criterion. For its calculation, we use the formula from (Mahmudov 2005):

$$PL = 69,55 + 26,16 \lg f_{mh} - 13,82 \lg h_1 - a(h_2) + \\ + (44,9 - 6,55 \lg h_1) \lg d + S \quad (1)$$

where

$$d = \sqrt{(\tau_1 - x_1)^2 + (\tau_2 - x_2)^2} , \quad (2)$$

$$a(h_2) = (1,1 \lg f_{mh} - 0,7) h_2 - (1,56 \lg f_{mh} - 0,8) , \quad (3)$$

$$S = 30 - 25 \lg B_l , \quad 0 < B_l < 51\% , \quad (4)$$

where f_{mh} is frequency, MHz; τ_1, τ_2 are location coordinates of BS; x_1, x_2 – location coordinates of the client; h_1 is the height of a BS antenna suspension, m; h_2 is the height of a receiver antenna, m; d is distance, *km*.

The formula is valid under the assumption that 15% of the area is built-up with buildings.

We apply these criteria of quality departing from:

1. The radio signal level depends on the distance between the transmitting station and the receiving station, but this dependency is nonlinear. Thus using the distance function as a criterion can give incorrect results.

2. Signal propagation, apart from the distance, depends on other parameters too, in particular on the altitude, at which a transmitting antenna is located. Formula (1) allows to take this factor into consideration.

Mathematical models can be formed in a discrete, continual or combined form.

2.2 Discrete problem statement

Supposing sets of cellular clients and possible locations of base stations are discrete. Then the optimal location problem can be described as follows:

Task 1.1. The set $J = \{1... m\}$ assigns possible locations of the base stations. In any of the items $j \in J$, one can install a base station and the value $b_j \geq 0$, $j = 1, 2, ... m$ is its capacity, namely – the maximum number of clients that it can serve simultaneously. The list of clients is defined by the set $I = \{1, ..., n\}$.

The task is to determine such N base stations location among possible positions that all clients would be connected to BS, limits for base station

capacity would be executed and the loss of radio signal would be minimized.

Let

$$x_{ij} = \begin{cases} 1, \text{ if the base station } j \text{ serves client } i, \\ 0, \text{ otherwise.} \end{cases} \quad (5)$$

$$s_j = \begin{cases} 1, \text{ if the base station is placed in position } j, \\ 0, \text{ otherwise.} \end{cases} \quad (6)$$

Using the introduced indications, the mathematical task model can be formulated as follows:

$$\max_{i \in I} \min_{j \in J} PL_{ij} \rightarrow \min , \quad (7)$$

$$\sum_{j=1}^{m} s_j = N, \quad (8)$$

$$\sum_{i=1}^{n} x_{ij} \leq b_j , \ j = 1,2, \ ... \ n , \quad (9)$$

$$\sum_{j=1}^{m} x_{ij} = 1, i = 1,2,...n . \quad (10)$$

Where PL_{ij} is the fading (loss in dB) of the signal, when the base station j serves the client i, calculated as follows:

$$PL_{ij} = 69,55 + 26,16 \lg f - 13,82 \lg h_1^j - a(h_2^i) + \\ + (44,9 - 6,55 \lg h_1^j) \lg d_{ij} + S \quad (11)$$

where

$$d_{ij} = \sqrt{(\tau_1^j - x_1^i)^2 + (\tau_2^j - x_2^i)^2} , \quad (12)$$

$$a(h_2^i) = (1,1 \lg f_{mh} - 0,7) h_2^i - (1,56 \lg f_{mh} - 0,8) \quad (13)$$

$$S = 30 - 25 \lg B_l , \quad 0 < B_l < 51\% \quad (14)$$

Task 1.2. Let the set $J = \{1, ..., m\}$ assign possible locations for the base stations. In any of the items $j \in J$ one can install the base station and the value $b_j \geq 0$, $j = 1, 2, ... m$ is its capacity, namely – the maximum number of clients it can serve simultaneously. The list of clients is defined by the set $I = \{1, ..., n\}$.

The task is to determine such N base stations location among possible positions that all clients would be connected and the pre-determined signal level P_{ij} would be provided, taking into account the power of each base station.

Using the indications introduced in Task 1.1, the mathematical model can be presented as follows:

$$\sum_{j=1}^{m} s_j \to \min, \tag{15}$$

$$\max_{i\in I} \min_{j\in J} PL_{ij} \le P , \tag{16}$$

$$\sum_{i=1}^{n} x_{ij} \le b_j , j = 1,2, \dots m, \tag{17}$$

$$\sum_{j=1}^{m} x_{ij} = 1, i = 1,2,\dots n. \tag{18}$$

Where PL_{ij} is fading (loss in dB) of the signal of the BS j serving client i, calculated by (11) – (14).

These tasks are the discrete optimal covering problems with additional restrictions of sets' power. For their solution the following methods can be used: greedy algorithm, statistically distributed greedy algorithm (Baldacci et al 2002); L-classes search algorithm (Kolokolov & Levanova 1996); decomposition algorithm (Bischoff et al 2009); genetic algorithms (Galenko 2002); a hybrid algorithm that is based on genetic algorithms and search strategies with restrictions (Francis & Lowe 1992), hybrid algorithm, developed on the basis of genetic algorithm and multistart strategy (Franci & Lowe 1992), algorithm based on lagrangian heuristics (Francis & Lowe 1992), and others (Bischoff et al 2009; Bischoff et al 2008).

However, the number of clients in actual problems is very large. Thus getting the solution of the problem in the problem statements 1.1 and 1.2 can be difficult due to complexity of the problem.

2.3 Continuous problem statement

Now let us consider these tasks in a continuous interpretation. Supposing cellphones consumers are located in a given area and a base station can be located in any place of this area. The problem of optimal location can be considered as follows:

Task 2.1. Supposing there is a set Ω of clients of mobile communication. They are placed in area Ω continually. The demand $\rho(x)$ for services at each point x of the set Ω is known. N base stations, with the set capacity b_j, $j = 1,2,\dots N$ should be placed in this area, and partition Ω_1,

Ω_2, ... Ω_N of the set Ω should be found, so that all clients would be connected to BS (each client is serviced by only one BS), limits for base station capacity would be executed and the loss of radio signal would be minimized.. Mathematical model of the stated problem can be presented as follows:

$$\max_{x\in\Omega_j} \min_{\tau\in\Omega} PL(x,\tau) \to \min_{(\tau^1,\dots\tau^N)\in\Omega^N} , \tag{19}$$

$$\bigcup_{j=1}^{N} \Omega_j = \Omega , \tag{20}$$

$$mes(\Omega_i \cap \Omega_j) = 0, i \ne j , \tag{21}$$

$$\int_{\Omega_j} \rho(x)dx \le b_j, j = \overline{1, N}. \tag{22}$$

Where

$$PL(x,\tau) = 69,55 + 26,16\lg f - 13,82\lg h_1^\tau - a(h_2^x) + \tag{23}$$
$$+ (44,9 - 6,55\lg h_1^\tau)\lg d(x,\tau) + S$$

$$d(x,\tau) = \sqrt{(\tau_1 - x_1)^2 + (\tau_2 - x_2)^2} , \tag{24}$$

$$a(h_2^x) = (1,1\lg f_{mh} - 0,7)h_2^x - (1,56\lg f_{mh} - 0,8) \tag{25}$$

$$S = 30 - 25\lg B_l , \quad 0 < B_l < 51\% \tag{26}$$

Task 2.2. Supposing there is a set Ω of the cellular clients. They are placed in area Ω continuously. The demand $\rho(x)$ for services at each point x of the set Ω is known. N base stations, with the set capacity b_j, $j = 1,2,\dots N$ should be placed in this area, and partition Ω_1, Ω_2, ... Ω_N of the set Ω should be found, so that all clients would be connected to BS (each client is serviced by only one BS), limits for base station capacity would be executed and the predetermined loss of radio signal would be achieved.

Let m be the number of installed base stations. Then the mathematical model of the problem can be presented as follows:

$$m \to \min, \tag{27}$$

$$\bigcup_{j=1}^{m} \Omega_j = \Omega, \tag{28}$$

$$mes(\Omega_i \cap \Omega_j) = 0, i \ne j, \tag{29}$$

$$\int_{\Omega_j} \rho(x)dx, j = \overline{1,N}, \qquad (30)$$

$$\max_{x \in \Omega} \min_{\tau \in \Omega} PL(x,\tau) \le P. \qquad (31)$$

Where $PL(x,\tau)$ is loss of signal in dB, when the base station τ serves the client x.

These tasks are analogous to continuous optimal coverage problem. The methods of its solution are considered in (Kiselova, Shor 2005; Goryachko & Us 2008; Kiselova et al 2009).

Taking into account that the base stations location is chosen out of some predetermined finite set, and the location of clients cannot be identified, because of their mobility, the model in which the set of possible locations of BS is discrete and customer set is continuous will be most appropriate. Thus, we arrive at the following problem statement.

2.4 Combined problem statement and algorithms

Task 3.1. Let the set $J = \{1, ..., m\}$ be the list of possible locations for the base stations. In any of the items $j \in J$ you can install a base station and value $b_j \ge 0$, $j = 1, 2, ... m$ is its capacity, namely – the maximum number of clients it can serve simultaneously. There is a set of mobile communication users, that are nonuniformly distributed in the area Ω, and the demand $\rho(x)$ for services at each point $x \in \Omega$ is known. The task is to determine the set $S \subseteq J, S \neq \Theta$ of the installed N base stations, and partition the set of clients Ω, so that all clients would be connected to BS, limits for base station capacity would be executed and the loss of the radio signal $PL(x,\tau)$ would be minimized.

Let Ω_j be the set of clients served by BS in position j,

$$s_j = \begin{cases} 1, \text{if the base station is placed in position j,} \\ 0, \text{otherwise.} \end{cases} \qquad (32)$$

Then the problem can be formulated as follows:

$$\max_{x \in \Omega} \min_{\tau \in S} PL(x,\tau) \to \min_{S \in J}, \qquad (33)$$

$$\sum_{j=1}^{m} s_j = N, \qquad (34)$$

$$\bigcup_{j=1}^{N} \Omega_j = \Omega, \qquad (35)$$

$$mes(\Omega_i \cap \Omega_j) = 0, i \neq j, \qquad (36)$$

$$\int_{\Omega_j} \rho(x)dx \le b_j, j = \overline{1,N}. \qquad (37)$$

Where $PL(x,\tau)$ is fading of the radio signal in dB (loss), when serving the client x.

The objective function in this model represents the maximal loss of radio signal. The constraint (34) means that we must install only N base stations; constraint (35) states that all clients are connected to BS; constraint (36) is equivalent to the requirement that each client is serviced by only one BS and constraint (37) ensures that limits for base station capacity are executed.

This problem is also analogous to the continuous problem of optimal coverage. Its solution can be obtained by the following algorithm based on the algorithm of the optimal set partition (Kiselova & Shor 2005),

Algorithm 3.1.

1. Choose subsets $S \in J$, $S = \{s_1, s_2, ...s_N\}$;

2. Find a partition $\Omega_1, \Omega_2, ... \Omega_N$, generated by a subset $S = \{s_1, s_2, ...s_N\}$ and satisfactory constraints (35) – (37);

2. Find the maximum signal fading PLj for each set $\Omega j, j = 1, 2, ... N$;

3. Choose the highest one among these values and take it as the radius of the coverage;

4. Of all the sets $S = \{s_1, s_2, ...s_N\}$ choose the one for which the radius is minimal.

End of algorithm.

Task 3.2. Let the set $J = \{1, ..., m\}$ be the list of possible locations for the base stations. In any of the items $j \in J$ you can install a base station and the value $b_j \ge 0$, $j = 1, 2, ... m$ is its capacity, namely – the maximum number of clients it can serve simultaneously. There is a set of mobile communication users, that are nonuniformly distributed in the area Ω, and the demand $\rho(x)$ for services at each point $x \in \Omega$ is known. The task is to determine a non vacuous set S of the installed base stations (BS) $S \subseteq J$ and partition the set of clients Ω on service areas of each station $\Omega_1, \Omega_2, ... \Omega_m$, so that all customers would be connected, limits for base station capacity would be executed and the predetermined radio signal level $PL(x,\tau)$ at every point of Ω would be maintained.

178

Let

$$s_j = \begin{cases} 1, \text{if the base station is placed in position } j, \\ 0, \text{ otherwise.} \end{cases} \quad (38)$$

The mathematical model of this problem that takes into account the considerations above, is:

$$\sum_{j=1}^{m} s_j \to \min, \quad (39)$$

$$\max_{x \in \Omega} \min_{\tau \in S} PL(x, \tau) \le P, \quad (40)$$

$$\bigcup_{j=1}^{m} \Omega_j = \Omega, \quad (41)$$

$$mes(\Omega_i \cap \Omega_j) = 0, i \ne j, \quad (42)$$

$$\int_{\Omega_j} \rho(x) dx \le b_j, j = \overline{1, m}. \quad (43)$$

Where $PL(x, \tau)$ is fading of the radio signal in dB, when servicing the client x. It was defined in the task 2.1.

Tasks 3.2 and Task 3.1 are related in some sense (Kiselova, Shor 2005). The solution to task 3.2 can be found with the help of following algorithm:

Algorithm 3.2.

1. Let us assume $N = 1$;

2. Select subsets $S_N = \{\tau_1, \tau_2, ...\tau_N\}, S \in J$.

3. Find a partition $\Omega_1, \Omega_2, ... \Omega_N$, generated by a subset $S = \{s_1, s_2, ...s_N\}$ and satisfactory constraints $(41) - (44)$, if it exists;

4. If the partition $\Omega_1, \Omega_2, ... \Omega_N$, generated by a subset $S = \{s_1, s_2, ...s_N\}$ does not exist, then set $N = N + 1$, and go to step 2,

5. Find the maximum fading of the radio signal PLj for each set $\Omega j, j = 1, 2, ... N$;

6. Choose the highest one among these values and take it as the radius of the coverage;

7. If the constraint (40) is not satisfied, then set $N = N + 1$, and go to step 2, otherwise go to step 8.

8. Among all sets S_N, which satisfy the constraint (40), choose the one which has minimum radius. It is the optimal solution.

End of the algorithm.

This algorithm is a modification of algorithm 3.1, consistently applied to $N = 1, 2, ... m$.

Let us also consider Task 3.3.

Task 3.3. Supposing there are some $S^1 = \{s_1, ..., s_M\}$ base stations at the given region, and their location is known. Let the set $J = \{1, ..., m\}$ be the set of possible locations for the base stations. In any of the items $j \in J$ one can install a base station and the value $b_j \ge 0$, $j = 1, 2, ... m$ is its capacity, namely – the maximum number of clients it can serve simultaneously. Users (clients) of mobile communication are nonuniformly distributed in area Ω, and the demand $\rho(x)$ for services at each point $x \in \Omega$ is known. The task is to determine a nonvacuous set S^2 of the installed base stations (BS) $S^2 \subseteq J$ and partition the set of clients Ω on service areas of each station $\Omega_1, \Omega_2, ... \Omega_m$, so that all clients would be serviced, limits for capacity would be executed and the predetermined radio signal level $PL(x, \tau)$ would be maintained.

Let M be the number of base station, which we already have. S is the set of all base stations (old and installed); Ω_j is the set of clients that are served by the base station j.

The mathematical model of this problem that takes into account all the above is as follows:

$$\sum_{j=1}^{m} s_j \to \min, \quad (44)$$

$$\max_{x \in \Omega} \min_{\tau \in S} PL(x, \tau) \le P, \quad (45)$$

$$\bigcup_{j=1}^{m} \Omega_j = \Omega, \quad (46)$$

$$mes(\Omega_i \cap \Omega_j) = 0, i \ne j, \quad (47)$$

$$\int_{\Omega_j} \rho(x) dx \le b_j, j = \overline{1, (N + m)}. \quad (48)$$

Where $PL(x, \tau)$ is fading of the radio signal in dB, when servicing the client x. It is defined in the same way as in tasks 3.1 and 3.2.

To solve this problem we can use the following algorithm, which is a modification of algorithm 3.2.

Algorithm 3.3.

1. Let us assume $N = 1$;

2. Select subsets

$S_{N+M} = \{\tau_1, \tau_2, ...\tau_N, \tau_{N+1}, \tau_{N+2}, ...\tau_{N+M}\}$,

$\{\tau_1, \tau_2, ...\tau_N\} \in J$, where the first N elements are fixed, and the other M elements are selected from J.

3. Find a partition $\Omega_1, \Omega_2, ... \Omega_{N+M}$ generated by a subset

$S_{N+M} = \{\tau_1, \tau_2, \ldots \tau_N, \tau_{N+1}, \tau_{N+2}, \ldots \tau_{N+M}\}$ and satisfactory constraints (46) – (48), if it exists;

4. If the partition $\Omega_1, \Omega_2, \ldots \Omega_{N+M}$, generated by a subset S_{N+M} does not exist, then set $N = N + 1$, and go to step 2,

5. Find the maximum signal fading PLj for each set $\Omega j, j = 1, 2, \ldots N+M$;

6. Choose the highest one among these values and take it as the radius of the coverage;

7. If the constraint (45) is not satisfied, then set $N = N + 1$, and go to step 2, otherwise go to step 8.

8. Among all sets S_{N+M}, which satisfy the constraint (46), choose the one which has the minimum radius. It is the optimal solution.

End of the algorithm.

We use the algorithm of optimal continual set partition to realize the step 3 of algorithms 3.1 – 3.3 and to find the partition $\Omega_1, \Omega_2, \ldots \Omega_N$, which satisfies the specified constraints. It is described in detail by E.M.Kiselova (Kiselova, Shor 2005) and is based on r-algorithm by N.Z. Shor (Shor 1985).

3 COMPUTATIONAL RESULTS

The proposed algorithms are numerically implemented and tested on the model tasks. Let us consider the results of applying the algorithm for solving the model task 3.2.

The input data for the model task are:
– the number of possible BS locations $N=25$ (coordinates of their locations and their height are shown in Table 1).
– Transmitter power is 38 dBm; antenna booster power is 14 dBm;
– $\Omega = \{(x_1, x_2) | 0 \le x_1 \le 200, 0 \le x_2 \le 200\}$;
– Limit of radio signal fading is 140.5 dBm.

Table 1. The coordinates of candidates for placement of base stations and their altitude

№	location coordinates		height, h (м)	№ of placement	location coordinates		height, h (м)	№ of placement	location coordinates		height, h (м)
	x_1	x_2			x_1	x_2			x_1	x_2	
1	20	35	45	9	70	75	42	17	70	140	54
2	18	180	39	10	113	86	42	18	65	170	48
3	98	196	54	11	140	70	42	19	22	135	45
4	180	20	48	12	186	99	42	20	145	160	54
5	63	15	51	13	46	41	45	21	50	174	48
6	95	50	45	14	85	110	51	22	85	179	42
7	133	28	54	15	120	120	42	23	118	186	39
8	32	71	45	16	148	114	57	24	182	191	45
								25	154	40	39

The results of algorithm application are shown in Table 2 and Figure 1. The required quality of signal was obtained when 5 base station had been located. Placement of more than 15 BS is excessive.

Table 2. The results of algorithm application

Iteration	Number of the BS	Signal loss
1	1	157,631
2	3	152,189
3	5	148,129
4	8	144,765
5	10	142,912
6	12	141,576
7	14	140,824
8	15	140,309
9	18	140,309

Figure 1. Optimal partition obtained for 15 base station

4 CONCLUSION

In this paper we considered the problem of base station location. Some of the appropriate mathematical models described in discrete, continuous and combined forms are proposed. The proposed combined model allows to consider clients' demand in any point of the region and take to account the limits of the base station capacity. The function of the signal fading considering not only the client's location, but also the height of antenna; was chosen as a criterion for the problem of optimal BS location, which allows to partition clients more accurately. The algorithm solution based on continual set partitioning method is proposed for the combined model. Experiments show that the proposed algorithm is feasible, gives the best results and can be successfully used.

REFERENCE

Mahmudov, M. 2005. *Improving consumer quality of telecommunications services by the example of a cellular mobile radio communications* (in Russian). InfoCOM.uz, Issue 5: 27 – 30.

Vishnevskij, V. 2003. *Theoretical bases of designing computer networks* (in Russian). Moscow: Texnosfera: 512.

Piyavskij, S. 1968. *On the optimization of networks* (in Russian). Proceedings of the Academy of Sciences of the USSR. Seria: Technical cybernetics, Issue 1: 68 – 80.

Galenko, S. 2002. *The use of hybrid algorithms for solving optimal location of base stations in the engineering of wireless data networks* (in Russian). http://www.masters.donntu.edu.ua/2007/kita/bukiy/ library/l3.htm

Baldacci, R., Hadjiconstantinou, E., Maniezzo, V. & Mingozzi, A. 2002. *A new method for solving capacitated location problem based on a set partitioning approach* . Computers and Operanions Research, Vol. 29: 365 –386.

Baldacci, R., Christofides, N. & Mingozzi, A. 2008. *An exact algorithm for the venicle routing problem based on a set partitioning formulation with additional cuts*. Mathematical Programming, Vol. 115, Issue 2: 351 –385.

Kiseleva, E. & Shor, N. 2005. *Continuous optimal set partitions: theory, algorithms, applications: Monograph* (in Russian). Kiev: Naukova dumka.

Kiseleva, E., Lozovskaja, L. & Timoshenko, E. 2009. *Using the theory of optimal partitioning of sets for solution of optival continuous coverage balls problem* (in Russian). Cybernetics and Systems Analysis, Issue 3: 98 –118.

Francis, R. & Lowe, T. 1992. *Aggregation error for location models: survey and analysis.* Annals of Operation Research, Vol. 40: 229 – 246.

Drezner, Z. & Hamacher, H. 2001. *Facility Location: Application and Theory*. Berlin: Springer: 457.

Beresnev, A. 2002. *Localization algorithm for user traffic in space and time in cellular of mobile radio communications systems of standard GSM* (in Russian). Journal of radioelectronics, Issue 11.

Kolokolov, A. & Levanova, T. 1996. *Decomposition algorithms and exhaustive search L-classes to solve some locatoin problems* (in Russian). Bulletin of Omsk University, Vol. 1: 21-23.

Bischoff, M., Fleischmann, T. & Klamroth, K. 2009. *The multi-facility locftion-allocftion problem with polyhedral barries.* Computers and Operanions Research, Vol. 36, Issue 5: 1376 –1392.

Brimberg, J., Hansen, P., Mladenovic, N. & Salhi, S. 2008. *A Survey of Solution Methods for the Continuous Location-Allocation Problem.* International Jornal of Operation Research, Vol. 5, Issue 1: 1 – 12.

Goryachko, E. & Us, S. 2008. *Mathematical modeling in the problems of optimal location of base stations in the engineering of wireless communication networks* (in Russian). Scientific Bulletin of National Mining University, Issue 5: 6 – 9.

Shor, N.Z. 1985. *Minimization methods for non-differentiable functions.* Berlin: Springer: 512.

Energy Efficiency Improvement of Geotechnical Systems – Pivnyak, Beshta & Alekseyev (eds)
© 2013 Taylor & Francis Group, London, ISBN 978-1-138-00126-8

Methods and principles of control over the complex objects of mining and metallurgical production

V. Kornienko, A. Gerasina & A. Gusev
State Higher Educational Institution "National Mining University", Dnipropetrovs'k, Ukraine

ABSTRACT: Methods and principles of control over the complex nonlinear technological objects are considered. The methods for estimation and prediction of the state of these objects are developed. A novel structurally-parametrical identification methodology based on the controlled object (CO) mathematical model is performed from the point of view of nonlinear dynamics. The method of synthesis is developed for optimal control of a nonlinear stochastic technological object combining the synergetic approach and the principle of a minimum of the generalized work.

1 INTRODUCTION

From the point of view of control, the complex controlled objects are the dynamic objects with non-stationary parameters, nonlinear dependences and stochastic variables. These include, for example, the technological processes of blast-furnace production and ore preparation (crushing and comminution) the expenses on which constitute a significant part of the cost of mining and metallurgical production (Kaganov, Blinov & Belen'kiy 1974; Maryuta, Kachan & Byn'ko 1983). Therefore, the task of increasing the efficiency of these processes through the creation of automatic control systems is extremely urgent.

2 FORMULATING THE PROBLEM

The basic conception of the modern automatic control theory consists (Krasovskiy 1987) in the achievement of main goals at each stage of functioning of the system, provided by the CO optimization in real time with the use of available a priori information at the stages of:

- estimation (filtration) of dynamic processes in COs;
- identification of the CO-model structure and parameters;
- the synthesis of optimal control by cycles of functioning of the system;
- adaptation (configuration of optimal control with incomplete information).

The synthesized model of a complex CO, which correctly transmits the dynamics of one mode of functioning, can be inadequate to the description of another one. Therefore, realization of the CO adaptive identification is needed in the process of functioning of the control system (CS).

At present, for complex nonlinear COs the intellectual control methods are being developed which consider an object not as an absolutely known point in the space of signs, but as some information about it. With such an approach, attempts are made to reproduce the principles of natural control systems – nervous systems of living organisms which realize the universal principles of processing of empiric information and search algorithms of adaptation (Kryglov, Dli & Golynov 2001).

In relation to the optimal control, one of the universal and effective (in the practical application to complex nonlinear systems) is the principle of a minimum of the generalized work, which is realized by means of the method with the predictive model and the functionals of the generalized work (FGW) (Krasovskiy 1987) and also the sinergistical principle of control (Kolesnikov 2000) which forms target modes for the solution of the tasks of optimal control over corresponding COs.

Thus, the purpose of this study is to develop and substantiate the methodology of constructing the algorithms for estimation, identification, synthesis of optimal control and adaptation for the creation of highly efficient systems of automatic control of nonlinear technological processes with intellectual prediction.

3 ESTIMATION AND PREDICTION OF CONTROL OBJECTS' STATE

In control theory, observation of some specified signal $q[k]$ with the less possible error is reduced to the problem of filtering.

1. As a rule, adaptive filter approximators (AFAs), having recursive and non-recursive connections, are used to realize the systems of estimation and identification of COs.

2. The flow diagrams of using the AFA as a predictive filter of signals and an approximator (of a standard model) of COs are given in Fig. 1.

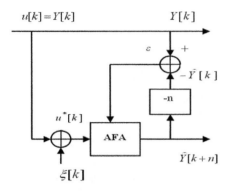
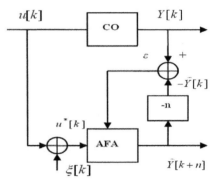

Figure 1. Prediction of a signal (a) and identification of a CO (b) by means of an AFA

Here delay and prediction of n steps marked as $-n$ and $+n$ and measured values of inputs are equal to $u*[k] = u[k] + \xi[k]$, where $\xi[k]$ is a noise measurement.

During operation of the AFA on each step by the error magnitude $\varepsilon[k] = Y[k] + \hat{Y}[k]$ between the actual $Y[k]$ and predicted $\hat{Y}[k]$ values of the signal (CO output) the adaptation of the AFA parameters is realized.

The differential equation of the predictive linear AFA with a finite impulse response (FIR) has the form:

$$\hat{Y}[k+n] = \sum_{r=0}^{R} b_r \cdot u[k-r] \qquad (1)$$

where b_r, R – coefficients of a filter and its order, respectively.

The differential equation of the linear AFA with infinite impulse response (IIR) has the following form:

$$\hat{Y}[k+n] = \sum_{m=1}^{M} a_m \cdot Y[k-m] + \sum_{r=0}^{R} b_r \cdot u[k-r] \qquad (2)$$

where a_m, M – coefficients of the filter and an order of its feedback, respectively.

In most adaptation algorithms for linear AFAs a square error ε is minimized by approximating the gradient of the objective function and iterative adjustment of filter coefficients.

For the construction and realization of the dynamic predictive model structure of a complex CO, the means of intellectual processing the information are preferably used as they are easily adjusted (adapted) to the changing CO properties, and, respectively, are effective means of modelling complex systems.

In terms of predictive systems on the basis of a neural network (NN), the best qualities are displayed by the heterogeneous network which consists of hidden layers with a nonlinear neuron activation function and an output linear neuron (Kryglov, Dli & Golynov 2001).

The CO model which predicts n steps on the basis of the NN of direct distribution with a hidden layer is represented in the form of a convolution equation:

$$\hat{Y}[k+n] = \sum_{\tau \in P} F_{\hat{x}} \{ \sum_{l \in Q} v_l[\tau] \cdot F_l (\sum_{m \in Q} v_{l,m}[\tau] \cdot u_m[k-\tau]) \} \qquad (3)$$

where P – a multitude of memory's depths of the corresponding inputs; $F_{\hat{x}}$ – an activation function of the NN output layer; Q – a set of neuron inputs; l – a serial number of an input of the NN output layer; v_l – weight coefficients of the output layer; F_l – an activation function of the hidden layer of neurons; m – a serial number of the NN input; $v_{l,m}$ – weight coefficients of the connection of m-input and l-neuron; u_m – a NN input.

The parameter settings (for training) of this NN are $\{v_l, v_{l,m}\} \subset a$.

Prediction by means of a NN with radial base functions (RBFs) is performed according to the equation:

$$\widehat{Y}[k+n] = \sum_{\tau \in P} F_{\bar{x}} \{ \sum_{l,m \in Q} v_l \cdot F_l(\vartheta_l, \|u_m[k-\tau] - v_l\|) \} \qquad (4)$$

where ϑ_l, v_l – parameters of the RBF of the hidden l-neuron layer.

Parameters of tuning of the NN (4) are $\{v_l, \vartheta_l, v_l\} \subset a$.

Identification of the NN parameters of direct distribution (3) is realized, as a rule, by means of gradient algorithms, for example, an algorithm of reverse distribution of an error in the space of parameters $\{v_l, v_{l,m}\} \subset a$ at a specified CO model to the structure (the NN architecture) and structural functions of F for the purpose of error prediction minimization.

When training a NN with a RBF (4), first centers and deviations for radial elements are determined, then parameters of a linear output layer are optimized: $\{v_l, \vartheta_l, v_l\} \subset a$.

Fuzzy logic systems and NNs are equivalent to each other, however, in practice they have their own advantages and disadvantages. In this connection, hybrid NNs came into existence in which conclusions are drawn on the basis of a fuzzy logic apparatus, and functions of belonging are adjusted with the use of teaching algorithms for training the NNs.

The adaptive neuro-fuzzy inference system (Anfis) is an example of a hybrid network (Kryglov, Dli & Golynov 2001) on the basis of which the CO equation has the form:

$$\widehat{Y}[k+n] = \sum_{\tau \in P} \sum_{m \in Q} \beta_m[\tau] \cdot \alpha_m[k-\tau] \qquad (5)$$

where $\qquad \beta_m[\tau] = U_m^{-1}(\alpha_m[\tau] / \sum_m \alpha_m[\tau]);$

$\alpha_m[k-\tau] = \underset{l,m \in Q}{Tn} \{ L_{l,m}(u_m[k-\tau]) \};$

$U = U(a_U); \quad L = L(a_L).$

Here U_m^{-1} – a function, reverse to the function of belonging of the intermediate output m of the

network with parameters a_U; α_m – a value of intermediate output; Tn – an arbitrary t-norm (Kryglov, Dli & Golynov 2001) of modelling a logical operation of "AND"; $L_{l,m}$ – the function of belonging of a fuzzy rule l of the input m with parameters a_L.

The parameters of tuning a NN (5) are $\{a_U, a_L\} \subset a$, its training is performed by analogy with a NN.

For example, in filters with transformations, in a frequency area as compared to convolution in a time domain, the volume of calculations is considerably reduced and the convergence properties of adaptation algorithms are improved.

The neuron wavelet (WVNN) AFA (Kornienko, Kyznetsov & Garnak 2009) is an example of such a filter. Here the process of filtering is based on the procedure of a direct discrete wavelet transformation (DWT). To diminish the influence of noise in the filter, the threshold limitation of coefficients of wavelet decomposition is realized and a NN is used to predict the values of coefficients according to which with the help of reverse DWT the predicted signal is determined.

4 IDENTIFICATION OF THE GENERATING PROCESSES

Quality control is directly connected with the accuracy of models of the processes occurring in the control system (CS).

Identification of a dynamic CO consists in the use of experimental data for obtaining or clarifying its mathematical model which sufficiently exactly (in a sense of the accepted criterion) approximates the CO in terms of input and output variables in the whole functional space (Krasovskiy 1987).

Identificanion of complex COs by traditional methods requires big expenses on experimental studies. The methods of nonlinear dynamics allow the operating modes of technological processes on separate time realizations to be determined (classified) and investigated and their models to be synthesized from common positions.

An experimental signal contains information about the mode of operations (attractor) of the generating system. It is proved (Kyznetsov 2002; Shyster 1988) that with one observed realization it is possible to determine the phase portrait of an attractor (mode of operation); the cross-correlation dimension D_C (the low boundary of Hausdorf dimention); the dimension of investment of an attractor d (dimension of a phase space) of the dynamic system, and also the correlation entropy K_C (the low boundary of entropy of Kolmogorov) which characterizes the signal randomness.

On the realization of the signal $x = x(t)$, by setting a delay τ and a dimension d of a phase space, its discrete reflection is built:

$$x[k] = \{x[k], x[k-m], x[k-2m], ..., \\ x[k-(d-1)m]\} \qquad (6)$$

where k – a step of time $t = k \cdot T$; T – a sampling period; m – delay steps ($m = \tau/T$); x – a vector of coordinates.

At the surplus on k, a discrete set of points is obtained in the d-dimensional space which under the permanent system mode and in accordance with the theorem of Takens (Takens 1980) is a phase portrait of the attractor.

The distance between the nearest points of the attractor before and after the bifurcations is in a universal relation. Self-similarity of such a phenomenon is described with the help of the fractal dimension of Hausdorf, and numerical determination of the attractor dimension is performed by means of a correlation dimension (Kyznetsov 2002; Shyster 1988):

$$D_C = \lim_{\varepsilon \to 0} \frac{\log\left(\sum\limits_{i=0}^{N(\varepsilon)} p_i^2\right)}{\log \varepsilon} = \lim_{\varepsilon \to 0} \frac{\log C(\varepsilon)}{\log \varepsilon},$$

$$C(\varepsilon) = \lim_{N \to \infty} N^{-2} \sum_{i,j} \chi[\varepsilon - \|x_i - x_j\|] \qquad (7)$$

where $C(\varepsilon)$ – a correlation integral estimated on the reflection of (6); N – duration of time realization; ε – size of a phase space cell; χ – Hevisayd's function.

From the investment theorem it follows (Shyster 1988) that the estimation of dimension of a phase space d is calculated through the estimation of the dimension of an attractor D_C of the real dynamic system (formula of Mane):

$$d \geq 2D_C + 1. \qquad (8)$$

The major characteristic of motion in the phase space of an arbitrary dimension is entropy of Kolmogorov K, which describes the dynamic behaviour of an attractor. Entropy of Kolmogorov K is proportional to the speed of loss of information on the state of a dynamic system in the course of time and shows the extent to which the dynamic system is chaotic.

The estimation (from below) of Kolmogorov's entropy K can be obtained with the account of (7) in the form (Kyznetsov 2002):

$$K_C = - \lim_{\varepsilon \to 0} \lim_{k \to \infty} \ln[C_k(\varepsilon)/C_{k+1}(\varepsilon)] \leq K,$$

$$C_k(\varepsilon) = \lim_{N \to \infty} N^{-2} \sum_{i,j} \chi[\varepsilon - \|x_i - x_j\|_k] \qquad (9)$$

where $\|x_i - x_j\|_k = \sqrt{\sum\limits_{n=0}^{k-1}(x_{i+n} - x_{j+n})^2}$.

K-entropy is equal to a zero for regular motion, endless – for random systems, positive and limited – for systems with the mode of chaos.

The exact prediction of the system's state is possible only on the interval of time T_C, as such $\varepsilon \cdot e^{KT_C} = 1$ (Kyznetsov 2002). Then the estimation (from above) of the exact prediction interval equals:

$$T_C = \frac{1}{K_C} \ln\left(\frac{1}{\varepsilon}\right). \qquad (10)$$

For the time, more than T_C, only statistic prediction is possible, the interval (depth) of which depends on the correlation function of the process.

Reconstruction of the dynamic system model consists in the choice of basic functions (structures) and their coefficients (parameters) of the model and also in the determination of the model parameter values, optimally corresponding to the time realization.

For the solution of the task of reconstructing the model CO, the d-dimentional reflection is formed:

$$x_{1,i+1} = \Phi_1\{x_{1,i}, x_{2,i}, ..., x_{d,i} \mid \lambda\};$$
$$........ \qquad (11)$$
$$x_{d,i+1} = \Phi_d\{x_{1,i}, x_{2,i}, ..., x_{d,i} \mid \lambda\},$$

where $x_{j,i}$ – the coordinates of the state vector in the moments of time $i \cdot T$; $j = \overline{1, d}$; $i = \overline{d, N-1}$.

Then the evolutional functions $\Phi_j\{x_i\}$ are represented as the decomposition on some basis with the sought after coefficients (parameters) λ .

The values of the λ parameters are chosen so that they can best meet the observed time realizations, for example, in accordance with the criterion of minimum error.

5 STRUCTURALLY-PARAMETRICAL IDENTIFICATION OF CONTROLLED OBJECTS

Let us formulate the task of identification of COs as follows: on the basis of an experimental set of functions (time series) of pertrubations, controls and outputs in the conditions of interferences to determine the structure (generalized function Φ) and vector of parameters a of the model:

$$\hat{Y}[k+n] = \Phi\{Y[k],u[k],w[k],\xi[k],a[k],k\} \quad (12)$$

that sufficiently exactly (in the sense of some criteria) approximate the COs in relation to the input and output values in the whole functional space. Here $Y[k],u[k],w[k],\xi[k]$ – respectively output vectors (matrices) of a CO, its controls, pertrubations and interferences to the current time k with the relevant memory's depths; n – a depth of prediction (for indemnification of clean delay and time for a synthesis and realization of control), corresponding to the interval of prediction ($n < T_C / T$).

Thus, formation of the vector $I_s = \{\Phi,a\}$ of estimating the structure Φ (structural identification) and parameters a (parametrical identification) of the CO model (12) is realized on the basis of signal vectors of observation by minimizing the accepted functional:

$$J[I_s] \to \min_{I_s \in S} J \Rightarrow I_s^{opt} = \{\Phi_{opt}, a_{opt}\} \quad (13)$$

where limitations S in a general case are equal to:

$$S = \begin{cases} \{h(I_s) \geq 0\} \subseteq S_h; \\ \{g(I_s) = 0\} \subseteq S_g; \\ \{\varphi_i, i = \overline{1,D}\} \subseteq S_D. \end{cases} \quad (14)$$

Here h, g – continuous functions; φ_i – elements of a discrete vector D of possible values of the structural function $\Phi = \{\varphi_i\}$.

Expressions (13)-(14) are a combination of a continuous task of mathematical and discrete programming which because of non-linearity of COs and an arbitrary type of the functional, are polymodals. And this requires the use of global optimization methods, among which searching methods are most effective (Takens 1980; Holland 1994; Rastrigin 1981). In them the algorithm for search of optimal solution links the solutions following each other $I_s(\kappa+1) = F[I_s(\kappa)]$, where F is a search algorithm which specifies what operations it is necessary to do on step κ at $I_s(\kappa)$ in order to obtain the solution $I_s(\kappa+1) \succ I_s(\kappa)$. Here the sign of advantage $\succ$ during minimization of the functional makes sense:

$$J[I_s(\kappa+1)] < J[I_s(\kappa)]. \quad (15)$$

Subsequent developments of these methods are genetic algorithms (GAs) (Holland 1994), based on the simulation of biological population development at the level of genomes. They simulate the process of biological evolution: mutations of the structure and parameters δI_s, their crossing (reproduction) $I_s(\kappa+1) = I_s(\kappa) + \delta I_s(\kappa)$ and rules of selection which allows us to find out their favourable variations by which a sequence of improved solutions with the property is built (15).

External criteria which are adequate to the task of constructing the models with minimum dispersion of a prediction error, and which are divided into the criteria of regularity and criteria of unbiasedness (a minimum of displacement) (Nelles 2001) are considered to be effective for structural identification. The criteria of regularity include the criterion of a minimum of relative error of incremental integration which is calculated on the entire sampling of the experimental data N , and also the criterion of a minimum of selective relative error at which optimization of the model is realized on the teaching sampling, and estimation of its efficiency – on the verification sampling.

When solving the task (13)-(14), the limitations $S_1 = S_h \cap S_g$ form the continuous task of mathematical programming and $S_2 = S_h \cap S_D$ – the task of discrete programming. In this case, the limitations S_1 can be taken into account when the step of search at which these limitations are performed is considered to be successful. With limitations S_2, the step of search is considered to be successful when the condition (15) is performed.

For the construction and realization of the structure of the dynamic predictive model of a CO (12) different approaches are used (Krasovskiy 1987). Thus, the nonlinear dynamic system (model of a CO) is represented as a composition of linear dynamic (LDL) and nonlinear static (NSL) links, for example, as models of Wiener, Gammerstain as well as their combination etc. In the quality of the LDLs the delay lines are used, the sizes of which (depths of memory) are determined by the dimension of input and output variables, and in the quality of the NSLs both traditional means: polynomials of Legendre or Kolmogorov-Gabor and intellectual ones: neuron networks (NN), hybrid NN with a fuzzy logic etc can be used.

6 ADAPTIVE CONTROL WITH PREDICTION

For the identification of a CO (12) investigation has found widespread use of searchless algorithms of parametrical identification with an adaptive model, oriented on the functioning in the real scale of time, to which gradient algorithms refer. Here the process of identification consists in adaptation of the parameters a on the functional error value J between the real output and the model response (by the functional gradient in a space of parameters):

$$\hat{a}[k] = \hat{a}[k-1] - K_a \cdot \nabla_{\hat{a}} J\{Y[k], \hat{Y}[k], k\} \quad (16)$$

where $\hat{a}[k]$ – estimation of the vector of adapted parameters on the current step; $\nabla_{\hat{a}} = (\partial / \partial \hat{a})^T$ – a gradient symbol; K_a – a

specified matrix of coefficients; $Y[k], \hat{Y}[k]$ – outputs of a CO and a model (12).

Usually the error functional J has the following form:

$$J_1 = \frac{1}{2} E\{(e[k])^2\} = \frac{1}{2} E\{(Y[k] - \hat{Y}[k])^2\} \quad (17)$$

where E – a mathematical expectation.

The AFA (a CO model) becomes optimal at $\hat{a} = a_{opt}$, when $J_1 = 0$, then the task of adaptation consists in the search of optimal coefficients $\hat{a}$ by the iterative determination of the gradient of the mean-squared error surface (17).

6.1 Self-tuning controller

According to the classification, this is an adaptive CS with the identification of a CO by tuning its model.

A CO and a REG form the basic contour of the system, and an AFA performs the role of a standard predictive model of the CO, which corrects coefficients of the regulator and is an informative contour. That it, this is a CS with open basic and closed information contours.

The synthesis of adaptive control in this CS on step k consists in:

– adaptations of coefficients of the AFA $\hat{a}[k]$ on the error value of the CO model (12) (for example, according to (16), on the functional gradient (17)), and then calculation of the output prediction of the CO on the adapted AFA $\hat{Y}[k+n]$;

– calculation according to the task on the next step of control $q[k+1]$ of the adjusting error

$$\varepsilon[k] = q[k+1] - \hat{Y}[k+n] \quad (18)$$

and determination of the control value $u[k]$ on its basis.

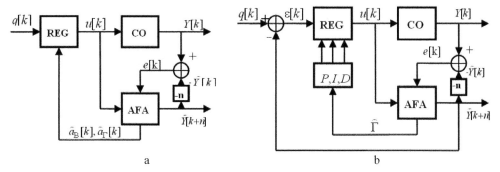

Figure 2. Structure of the CS with a self-tuning controller (a) and of the CS with a PID-controller (b) and with the prediction

The purpose of the adjustment is providing a zero error

$$\varepsilon^*[k] = q[k+1] - Y[k+n] = 0 .$$

As the signal $Y[k+n]$ is not observed in the moment of time k, its estimation (prediction) on the AFA $\hat{Y}[k+n]$ is used, that meets the criterion of adjustment for the stochastic CO:

$$J_2 = \frac{1}{2} E\{(\varepsilon[k])^2\} =$$
$$= \frac{1}{2} E\{(q[k+1] - \hat{Y}[k+n])^2\}$$
(19)

Then the control value $u[k]$ is determined from the equation:

$$E\{(q[k+1] - \Phi\{Y[k], u[k], \xi[k], a[k], k\})^2\} = 0$$
(20)

that meets the functional minimum (19) and is the task of minimizing the function of one variable.

Let the model CO (12) be represented as a linear control equation:

$$\hat{Y}[k+n] = B\{Y[k], \hat{a}_B[k]\} + \Gamma\{Y[k], \hat{a}_\Gamma[k]\} \times \\ \times u[k]$$
(21)

where B – a function of autonomous motion of a CO; Γ – a function of control sensitiveness; $\hat{a}_B[k]$, $\hat{a}_\Gamma[k]$ – vectors of tuning parameters of the functions B and Γ .

If the parameters $\hat{a}_B[k]$, $\hat{a}_\Gamma[k]$ are determined, the equation of a regulator is equal to:

$$u[k] = \frac{q[k+1] - B\{Y[k], \hat{a}_B[k]\}}{\Gamma\{Y[k], \hat{a}_\Gamma[k]\}} .$$
(22)

6.2 PID-controller

The structure of an adaptive CS with a PID-controller and an intellectual predictive model as an AFA is presented in Fig. 2, b. It is a closed CS by its basic and information contours.

The synthesis of adaptive control in this CS also includes the prediction calculation procedure $\hat{Y}[k+n]$ on the adapted model of a CO, which is analogical to the one considered above, and the procedure for determining the control action $u[k]$, which has certain features.

The equation of a PID-controller is represented by a discrete equation:

$$u[k] = u[k-1] + P[k] \cdot (\varepsilon[k] - \varepsilon[k-1]) + \\ + I[k] \cdot \varepsilon[k] + D[k] \cdot (\varepsilon[k] - 2\varepsilon[k-1] + \varepsilon[k-2])$$
(23)

where $\varepsilon[k]$ – a regulation error (18), $\Theta = \{P, I, D\}$ – amplification coefficients of PID-channels.

The procedure of determining the control action $u[k]$ in this CS consists in the search of parameter values of Θ (23) which are the solutions to the equation (20) and meet functional minimum (19). It is a task of minimization of function of a few variables.

When meeting the requirements for the use of gradient methods of the functional minimum (19) with the account of (23) corresponds:

$$\partial J_2[k] / \partial \Theta[k] = 0 ;$$
$$\Theta[k] = \{P[k], I[k], D[k]\}$$
(24)

from which current parameter values $\Theta[k]$ are determined analogically to the expression (16):

$$\Theta[k] = \Theta[k-1] - \mu_\Theta \cdot \partial J_2[k] / \partial \Theta[k] \quad (25)$$

where $\mu_\Theta = \{\mu_P, \mu_I, \mu_D\}$ are coefficients of tuning of PID-channels.

For the models of COs with linear incoming control (21) in accordance with (23) the equation (25) takes the form:

$$P[k] = P[k-1] + \mu_P \cdot \varepsilon[k] \cdot \Gamma\{Y[k], \hat{a}_\Gamma[k]\} \div \quad (26)$$
$$\times (\varepsilon[k] - \varepsilon[k-1])$$

$$I[k] = I[k-1] + \mu_I \cdot \varepsilon[k] \cdot \Gamma\{Y[k], \hat{a}_\Gamma[k]\} \times \quad (27)$$
$$\times \varepsilon[k]$$

$$D[k] = D[k-1] + \mu_D \cdot \varepsilon[k] \cdot \Gamma\{Y[k], \hat{a}_\Gamma[k]\} \times \quad (28)$$
$$\times (\varepsilon[k] - 2\varepsilon[k-1] + \varepsilon[k-2])$$

7 OPTIMAL CONTROL OF NONLINEAR OBJECTS

7.1 Principle of a minimum of the generalized work

The task of a synthesis of optimal control of nonlinear dynamic COs in accordance with the principle of a minimum of the generalized work is solved as follows. Let the prediction model of a CO have a linear incoming control:

$$x[k+1] = \varphi(x[k]) + \psi(x[k]) \cdot u[k] + \xi[k] \quad (29)$$

where in accordance with (12) $\{Y[k], w[k]\} \subset \{x[k]\}$.

A synthesis is performed, for example, on the stochastic FGW with additive functions of expenses on the control and discrete times:

$$J_1 = E\{V_3(x[k_{j+1}]) + \sum_{k=k_j}^{k_{j+1}-1} Q_3(x[k], k) + \quad (30)$$

$$+ \sum_{k=k_j}^{k_{j+1}-1} U_3(u[k], k) + \sum_{k=k_j}^{k_{j+1}-1} U_3^*(u_{opt}[k], k)\}$$

where E – a mathematical expectation; V_3 – a terminal function of the final state of the control stage (the objective function); Q_3, U_3 – positively determined functions of expenses; U_3^* – a positively determined function which takes a minimum value at $u = u_{opt}$; u_{opt} – a required optimal control; k_j, k_{j+1} – initial steps of the successive stages of control.

A discrete equation of Bellman can be written as:

$$V_i(x[i]) = E\{Q_3(x[i]) + V_{i+1}\{\varphi(x[i]) \cdot u_{opt}[i]\} - $$
$$- \frac{\partial V_{i+1}}{\partial x[i+1]} \cdot \psi(x[i]) \cdot u_{opt}[i]\}$$
$$i = k_{j+1} - 1, k_{j+1} - 2, \dots \quad (31)$$

At the limiting condition $V_{k_{j+1}}(x[k_{j+1}]) = V_3(x[k_{j+1}])$ the solution of the recurrence equation (31) is determined in accordance with the equality:

$$\frac{\partial}{\partial u_{opt}} U_3(u_{opt}[i]) = -\frac{\partial V_{i+1}}{\partial x[i+1]} \cdot \psi(x[i]) \quad (32)$$

and, for example, when the quadratic function of control costs is equal to:

$$u_{opt}[i] = -K \cdot \frac{\partial V_{i+1}}{\partial x[i+1]} \cdot \psi(x[i]), \quad (33)$$

where K – a matrix of the specified positive coefficients.

For the algorithms of optimal control synthesis on the FGW with a predictive model, the solution of the equations (31), (32) is performed by the method of characteristics, the feature of which is the connection of the optimal solution with the free motion of a CO (at $u = 0$) on the predictable cycle of control $[k_j, k_{j+1}]$.

Optimal control on the FGW enables the construction of asymptotically steady (at $u = 0$) control systems, and the basic functional equation of a nonlinear CO is by itself a linear differential equation with partial derivatives which has simple numeral solutions.

7.2 Synergetic principle of optimal control

A sinergetic principle consists in the target method of self-organization of synthesized control systems where the purpose as an attractor (desired mode of operation) determines self-control and directed self-organization of a nonlinear dynamic process.

For the realization of a synergetic approach in control systems it is necessary to pass from the separate consideration of CO models and external forces (as controls, tasks and perturbations) to the extended target setting so that the marked forces became internal interactions of the general (closed) system.

The synthesis of optimal control consists in the determination of the control law which provides the convergence of the CO's state (29) with the desired mode of operation (by an invariant diversity) $\Psi(x[k]) = 0$ and further asymptotic steady motion of the CO along it to the beginning of co-ordinates of a phase space.

The basic functional equation of Euler-Lagrange

$$T\frac{\partial \Psi}{\partial x}\cdot\psi(x)\cdot u + T\frac{\partial \Psi}{\partial x}\cdot\varphi(x) + \Psi = 0 \qquad (34)$$

determines the control law

$$u[k+1] = -\left(\frac{\partial \Psi[k+1]}{\partial x[k]}\cdot\psi(x[k])\right)^{-1} \times$$

$$\times\left[\frac{\partial \Psi[k+1]}{\partial x[k]}\cdot\varphi(x[k]) + T^{-1}\cdot\phi(\Psi[k])\right] \qquad (35)$$

which delivers a minimum of the accompanying functional

$$J_2 = E\{\sum_{k=k_j}^{k_{j+1}-1}\{T^2\Psi^2[k+1] + \phi^2(\Psi[k])\}\}, \qquad (36)$$

where T – a matrix of positively determined coefficients; ϕ – a differentiated function which characterizes the required quality of the CO transient processes.

Thus, the nonlinear dynamics of the CO in the space of states is approximated by a linear dynamics in the space of macrovariables. And the task of selecting Ψ is reduced to the task of a synthesis of the steady homogeneous system of differential equations.

The considered algorithms ensure the synthesis of optimal control in the process of functioning of CSs and use the predictive models of COs. They include the following stages:

1) estimation of the current state of COs in moments of the beginning of the current cycle of control (k_j);

2) prediction of a free motion of COs on the model (12) at a given interval $[k_j, k_{j+1}]$ of control optimization;

3) calculation of the gradient of the objective function $V_{i+1}(x[i+1])$ or aggregated function $\Psi(x[k+1])$ for the current state of a CO;

4) formation of a signal of optimum control according to (33) or (35).

8 CONCLUSIONS

The performed research substantiates a novel methodology of constructing the control systems by means of complex technological processes on the basis of intellectual prediction. Such an approach allows us to reduce the development costs and improve the quality of CSs as it doesn't require the creation of exact CO models at the stage of planning. The control synthesis is realized on the intellectual forecasting models adapted in the process of CS functioning.

For the first time the use of the principle of a minimum of generalized work has been grounded for the technological objects of the mining and metallurgical production.

It has been shown that the optimal control on the FGW enables the construction of steady asymptotic control systems and that the basic functional equation of a nonlinear CO is by itself a linear differential equation with partial derivatives which has simple numerical solutions. In addition, the synergetic approach provides approximation of the CO nonlinear dynamics in the space of states with the linear dynamics in the space of macrovariables, the task of which is reduced to the synthesis of a steady homogeneous system of differential equations.

REFERENCES

Kaganov, V.Yu., Blinov, O.M. & Belen'kiy, A.M. 1974. *Automating the management of metallurgical processes* (in Russian). Moscow: Metallurgy: 415.

Maryuta, A.N., Kachan, Yu.G. & Byn'ko, V.A. 1983. *Automatic control of technological processes of processing plants* (in Russian). Moscow: Nedra: 277.

Krasovskiy, A.A. 1987. *Guide to the Theory of Automatic Control* (in Russian). Moscow: Nayka: 711.

Kryglov, V.V., Dli, M.I. & Golynov, R.Yu. 2001. *Fuzzy logic and artificial neural networks* (in Russian). Moscow: Fizmatlit: 221.

Kolesnikov, A.A. 2000. *Modern Applied Control Theory: The synergetic approach to management theory* (in Russian). Taganrog: Publishing house TRTU.

Kornienko, V.I., Kyznetsov, G.V. & Garnak, I.V. 2009. *Wavelet neural prediction and identification of complex signals and control objects* (in Ukrainian). Information technology and computer engineering, Issue 2 (15).

Kyznetsov, S.P. 2002. *Dynamic chaos* (in Russian). Moscow: Fizmatlit: 296.

Shyster, G. 1988. *Deterministic chaos. Introduction* (in Russian). Moscow: Mir: 240.

Takens, F. 1980. *Lecture Notes in Mathematics.* Berlin: Springer-Verlag: 898.

Holland, J.H. 1994. *Adaptation in natural and artificial systems. An introductory analysis with application to biology, control and artificial intelligence.* London: Bradford book edition.

Rastrigin, L.A. 1981. *The adaptation of complex systems* (in Russian). Riga: Zinatne: 375.

Nelles, O. 2001. *Nonlinear System Identification: From Classical Approaches to Neural and Fuzzy Models.* Berlin: Springer.

Energy Efficiency Improvement of Geotechnical Systems – Pivnyak, Beshta & Alekseyev (eds)
© 2013 Taylor & Francis Group, London, ISBN 978-1-138-00126-8

Intellectual support of making decisions under the conditions of indeterminacy in the process of control automation of drilling facilities

L. Meshcheriakov, L. Tokar & I. Udovyk
State Higher Educational Institution "National Mining University", Dnipropetrovs'k, Ukraine

ABSTRACT: The results of applying a method of intellectual identification of operative states of drilling facilities in the process of control automation through forming intellectual support of decision-making under the conditions of their operation modes real uncertainty are described.

INTRODUCTION

Under complicated conditions of indeterminacy concerning operational status of drilling facilities (DF) or its certain components as well as destructive properties of rocks, and depreciation of rock-breaking tool (RBT) while drilling, the process efficiency depends on capabilities and accuracy of differential technological and technical diagnostics; that is a classification of one of above-mentioned practical levels of complexity (Meshcheriakov 2006). In this case, quality of the diagnostics relies heavily on the DF personnel qualification (Dudlia et al. 2005). That stipulates urgency of software availability for intellectual support of making diagnostic decisions while drilling facilities controlling.

MAIN PART

According to control sampling of mode and technological parameters of RBT-type diamond crowns, residual life may be identified through diagnostic types and parameters of drilling facilities state at tracking levels (from the trough to the peak): x_1 – residual life of RBT is 5 m; x_2 – residual life of RBT is 10 m; x_3 – residual life of RBT is 15 m; x_4 – residual life of RBT is 20 m; and x_5 – residual life of RBT is 25 m.

The residual levels of RBT (x_1-x_5) are taken as diagnostic types to be recognized. While determining diagnosis of residual life of diamond crown for specific drilling facilities, following basic mode and technological parameters are taken into account being measured in the process of drilling mode (possible range of used parameter variations is parenthesized): y_1 is crown load H (800-1200 kgf); y_2 is rotational frequency of the crown ω (13.6-20 c^{-1}); y_3 is circulation rate Q (25-100 liters per minute); y_4 is drilling speed V (0.24-0.99 meters per hour); y_5 is sink rate per rotation h (0,044-0,108 mm per rotation); y_6 is drilling depth L (1000-2000 m); y_7 is drilling time T (2-30 hours); y_8 is trip time T_c (1-30 hours); y_9 is G rock category (8-10 c.u.); y_{10} is diamond number k (7,77-14,6 karat); y_{11} is single-point diamond diameter D (1,55-1,9 mm); and y_{12} is the diamond price S (47-90UAH).

In this case, the problem of function test lies in the fact that each parameter combination should be in accordance with one of $x_j(j = \overline{1,5})$ solutions. In this context, fuzzy data base is formed. Within the base, functionally defined basic parameters $y_1 - y_{12}$ are considered as linguistic variables in accordance with intellectual technology of identification (Meshcheriakov 2006). In addition, it is required to put extra linguistic variables in deduction tree (Figure 1): x is prognosticated resource

193

identified through $x_1 - x_5$ resource levels; z_1 are technological effects on the resource depending upon the set of mode parameters $\{y_1, y_2, y_3, y_4\}$; and z_2 are technological effects of the resource depending upon the set of mode parameters $\{y_5, y_6, y_7, y_8, y_9, y_{10}, y_{11}\}$.

A structure of a model for differential diagnostics of RBT-type diamond crowns of drilling facilities represented in the form of deduction (Figure 1) fits the equations:

$$x = f_x(z_1, z_2, y_{12});$$

$$z_1 = f_{z1}(y_1, y_2, y_3, y_4); \tag{1}$$

$$z_2 = f_{z2}(y_5, y_6, y_7, y_8, y_9, y_{10}, y_{11}).$$

To estimate values of $y_1 - y_{12}$, and z_1 i z_2 linguistic variables, the standard range of qualitative terms of intellectual technology of identification is applied: H – low, нC –below average, C – average, вC – above average, and B – high. Each of them is a fuzzy set stipulating and defining with the help of proper membership functions. With the help of introduced qualitative terms and drilling expert knowledge [2], one can represent formed ratios 1 in the form of Tables 2-4 of qualitative correlation of deduction tree of differential diagnostics of RBT-type diamond crown.

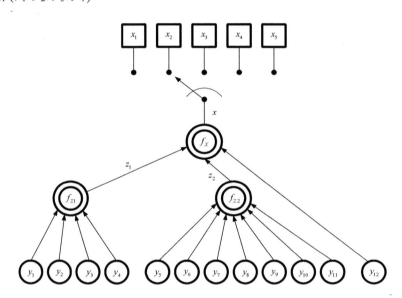

Figure 1. Deduction tree of RBT-type diamond crown

Table 1. Correlations of number values of basic technological and technical parameters of drilling by means of 01А3Д20К30 – 14А4Д50К50, D=76 mm diamond crowns and standard range of qualitative terms

Basic parameters	H	нC	C	вC	B
y_1	800 – 880	881 – 961	962 – 1042	1043 – 1123	1124 – 1200
y_2	13,6 – 15,8	15,9 – 18,1	18,2 – 20,4	20,5 – 22,7	22,8 – 25
y_3	25 – 40	41 – 55	56 – 70	71 – 85	86 – 100
y_4	0,24 – 0,38	0,39 – 0,53	0,54 – 0,68	0,69 – 0,83	0,84 – 0,99
y_5	0,029– 0,044	0,045– 0,060	0,061– 0,076	0,077– 0,092	0,093– 0,108

Basic parameters	H	нC	C	вC	B
y_6	$1000 - 1180$	$1181 - 1360$	$1361 - 1540$	$1541 - 1710$	$1711 - 1900$
y_7	$2 - 7$	$8 - 13$	$14 - 19$	$20 - 25$	$26 - 30$
y_8	$1 - 7$	$8 - 13$	$14 - 19$	$20 - 25$	$26 - 30$
y_9	$8 - 8,4$	$8,4 - 8,8$	$8,8 - 9,2$	$9,2 - 9,6$	$9,6 - 10$
y_{10}	$7,77 - 9$	$9 - 10,4$	$10,4 - 11,8$	$11,8 - 13,2$	$13,2 - 14,6$
y_{11}	$1,55 - 1,62$	$1,63 - 1,7$	$1,71 - 1,78$	$1,78 - 1,85$	$1,85 - 1,9$
y_{12}	$47 - 54$	$55 - 63$	$64 - 72$	$73 - 81$	$82 - 90$

In this context linguistic variable z_1 depending upon information concerning operative technological state of DF in turn forms following mode technological variables (ranges of the parameters variations are in parenthesis): y_1 is load on H crown (800-1200 kgf); y_2 is a crown rotational speed ω (13,6-20 c^{-1}); y_3 is circulation rate Q (25-100 liters per minute); and y_4 is drilling speed V (0,24-0,99 meters per minute) (Table 3).

Table 2. Ratios between z_1, z_2, y_{12} linguistic variables and qualitative terms of RBT diamond crowns of 01А3Д20К30 – 14А4Д50К50, D=76 mm types

z_1	z_2	y_{12}	x
H	H	H	
H	H	нC	x_1
нC	H	нC	
нC	нC	нC	
C	нC	нC	x_2
нC	нC	H	
C	C	C	
вC	C	C	x_3
вC	C	H	
вC	B	нC	
вC	вC	нC	x_4
B	вC	нC	
B	B	нC	
B	вC	нC	x_5
вC	B	нC	

Table 3. Technical ratios of linguistic variables $y_1 - y_4$ and qualitative terms of RBT diagnosis of 01А3Д20К30 – 14А4Д50К50, D=76 mm types

y_1	y_2	y_3	y_4	z_1
вС	Н	нС	нС	Н
В	Н	нС	Н	
С	Н	нС	Н	
вС	Н	Н	С	нС
В	Н	С	С	
В	Н	нС	С	
вС	Н	Н	вС	С
вС	Н	Н	В	
вС	Н	Н	вС	
С	В	С	В	вС
С	В	С	В	
С	В	С	вС	
вС	В	С	В	В
В	вС	вС	С	
вС	В	С	В	

Linguistic variable z_2 with its parameters is shown in Figure 4.

In turn linguistic variable z_2 depends on information concerning technical state of DF; adequate qualitative terms are formed through qualitative terms of technical variables: y_5 is depth per rotation h (0,044-0,108 mm per rotation); y_6 is drilling depth L (1000-2000 m); y_7 is drilling time T (2-30 hours); y_8 is T_c operations trip (1-30 hours); y_9 is G rocks category (8-10 c.u.); y_{10} the number of diamonds k (7,77-14,6 karat); y_{11} is single-point diamond diameter D (1,55-1,9 mm).

Table 4. Technical ratios of linguistic variables $y_5 - y_{11}$ and qualitative terms of RBT diagnosis of 01А3Д20К30 – 14А4Д50К50, D=76 mm types

y_5	y_6	y_7	y_8	y_9	y_{10}	y_{11}	z_2
Н	С	нС	Н	В	нС	Н	Н
Н	вС	Н	Н	С	нС	Н	
Н	нС	нС	Н	В	нС	Н	
С	вС	С	нС	С	Н	Н	нС
С	нС	С	Н	С	нС	Н	

196

y_5	y_6	y_7	y_8	y_9	y_{10}	y_{11}	z_2
C	вС	C	нС	C	нС	H	нС
B	вС	C	вС	C	C	вС	С
B	вС	C	C	C	C	вС	
B	вС	C	C	C	H	H	
C	H	C	B	B	нС	H	вС
C	H	C	B	C	нС	H	
C	H	C	B	C	нС	H	
C	H	B	B	C	нС	H	B
нС	вС	B	B	C	нС	H	
C	H	B	B	C	нС	H	

Systems of logic fuzzy equations connecting membership functions of diagnosis and input linguistic variables as well as qualitative terms are recorded with the help of formalized information concerning ratios between linguistic variables and qualitative terms of RBT diamond crowns of 01А3Д20К30 – 14А4Д50К50, D=76 mm type (Tables 2-4), and « $*$ » (And – min) and « $\vee$ » (OR – max) operations. As for RBT residual resource $x_1 - x_5$ diagnosis, a system of logical fuzzy equation is

$$v^{x_1}(x) = \left[v^H(z_1) \cdot v^H(z_2) \cdot v^H(y_{12}) \right] \vee \left[v^H(z_1) \cdot v^H(z_2) \cdot v^{HC}(y_{12}) \right] \vee \left[v^{HC}(z_1) \cdot v^H(z_2) \cdot v^{HC}(y_{12}) \right];$$

$$v^{x_2}(x) = \left[v^{HC}(z_1) \cdot v^{HC}(z_2) \cdot v^{HC}(y_{12}) \right] \vee \left[v^C(z_1) \cdot v^{HC}(z_2) \cdot v^{HC}(y_{12}) \right] \vee \left[v^{HC}(z_1) \cdot v^{HC}(z_2) \cdot v^H(y_{12}) \right];$$

$$v^{x_3}(x) = \left[v^C(z_1) \cdot v^C(z_2) \cdot v^C(y_{12}) \right] \vee \left[v^{6C}(z_1) \cdot v^C(z_2) \cdot v^C(y_{12}) \right] \vee \left[v^{6C}(z_1) \cdot v^C(z_2) \cdot v^H(y_{12}) \right]; \qquad (2)$$

$$v^{x_4}(x) = \left[v^{6C}(z_1) \cdot v^B(z_2) \cdot v^{HC}(y_{12}) \right] \vee \left[v^{6C}(z_1) \cdot v^{6C}(z_2) \cdot v^{HC}(y_{12}) \right] \vee \left[v^B(z_1) \cdot v^{6C}(z_2) \cdot v^{HC}(y_{12}) \right];$$

$$v^{x_5}(x) = \left[v^B(z_1) \cdot v^B(z_2) \cdot v^{HC}(y_{12}) \right] \vee \left[v^B(z_1) \cdot v^{6C}(z_2) \cdot v^{HC}(y_{12}) \right] \vee \left[v^{6C}(z_1) \cdot v^B(z_2) \cdot v^{HC}(y_{12}) \right].$$

Logical fuzzy equations system showing information on linguistic variable $y_1 - y_4$ - qualitative term ratios according to introduced variable z_1 of DF technological state is:

$$v^H(z_1) = \left[v^{6C}(y_1) \cdot v^H(y_2) \cdot v^{HC}(y_3) \cdot v^{HC}(y_4) \right] \vee \left[v^B(y_1) \cdot v^H(y_2) \cdot v^{HC}(y_3) \cdot v^H(y_4) \right] \vee$$
$$\vee \left[v^C(y_1) \cdot v^H(y_2) \cdot v^{HC}(y_3) \cdot v^H(y_4) \right];$$

$$v^{HC}(z_1) = \left[v^{6C}(y_1) \cdot v^H(y_2) \cdot v^H(y_3) \cdot v^C(y_4) \right] \vee \left[v^B(y_1) \cdot v^H(y_2) \cdot v^C(y_3) \cdot v^C(y_4) \right] \vee$$
$$\vee \left[v^B(y_1) \cdot v^H(y_2) \cdot v^{HC}(y_3) \cdot v^C(y_4) \right];$$

197

$$v^{C}(z_{1}) = \left[v^{6C}(y_{1}) \cdot v^{H}(y_{2}) \cdot v^{H}(y_{3}) \cdot v^{6C}(y_{4}) \right] \vee \left[v^{6C}(y_{1}) \cdot v^{H}(y_{2}) \cdot v^{H}(y_{3}) \cdot v^{B}(y_{4}) \right] \vee \tag{3}$$
$$\vee \left[v^{6C}(y_{1}) \cdot v^{H}(y_{2}) \cdot v^{H}(y_{3}) \cdot v^{6C}(y_{4}) \right];$$

$$v^{6C}(z_{1}) = \left[v^{C}(y_{1}) \cdot v^{B}(y_{2}) \cdot v^{C}(y_{3}) \cdot v^{B}(y_{4}) \right] \vee \left[v^{C}(y_{1}) \cdot v^{B}(y_{2}) \cdot v^{C}(y_{3}) \cdot v^{B}(y_{4}) \right] \vee$$
$$\vee \left[v^{C}(y_{1}) \cdot v^{B}(y_{2}) \cdot v^{C}(y_{3}) \cdot v^{6C}(y_{4}) \right];$$

$$v^{B}(z_{1}) = \left[v^{6C}(y_{1}) \cdot v^{B}(y_{2}) \cdot v^{C}(y_{3}) \cdot v^{B}(y_{4}) \right] \vee \left[v^{B}(y_{1}) \cdot v^{6C}(y_{2}) \cdot v^{6C}(y_{3}) \cdot v^{C}(y_{4}) \right] \vee$$
$$\vee \left[v^{6C}(y_{1}) \cdot v^{B}(y_{2}) \cdot v^{C}(y_{3}) \cdot v^{B}(y_{4}) \right].$$

Logical fuzzy equations system showing qualitative term ratios according to introduced information on linguistic variable $y_{5} - y_{11}$ - variable z_{2} of DF technological state is:

$$v^{H}(z_{2}) = \left[v^{H}(y_{5}) \cdot v^{C}(y_{6}) \cdot v^{HC}(y_{7}) \cdot v^{H}(y_{8}) \cdot v^{B}(y_{9}) \cdot v^{HC}(y_{10}) \cdot v^{H}(y_{11}) \right] \vee$$
$$\vee \left[v^{H}(y_{5}) \cdot v^{6C}(y_{6}) \cdot v^{H}(y_{7}) \cdot v^{H}(y_{8}) \cdot v^{C}(y_{9}) \cdot v^{HC}(y_{10}) \cdot v^{H}(y_{11}) \right] \vee$$
$$\vee \left[v^{H}(y_{5}) \cdot v^{HC}(y_{6}) \cdot v^{HC}(y_{7}) \cdot v^{H}(y_{8}) \cdot v^{B}(y_{9}) \cdot v^{HC}(y_{10}) \cdot v^{H}(y_{11}) \right];$$

$$v^{HC}(z_{2}) = \left[v^{C}(y_{5}) \cdot v^{6C}(y_{6}) \cdot v^{C}(y_{7}) \cdot v^{HC}(y_{8}) \cdot v^{C}(y_{9}) \cdot v^{H}(y_{10}) \cdot v^{H}(y_{11}) \right] \vee$$
$$\vee \left[v^{C}(y_{5}) \cdot v^{HC}(y_{6}) \cdot v^{C}(y_{7}) \cdot v^{H}(y_{8}) \cdot v^{C}(y_{9}) \cdot v^{HC}(y_{10}) \cdot v^{H}(y_{11}) \right] \vee$$
$$\vee \left[v^{C}(y_{5}) \cdot v^{6C}(y_{6}) \cdot v^{C}(y_{7}) \cdot v^{HC}(y_{8}) \cdot v^{C}(y_{9}) \cdot v^{HC}(y_{10}) \cdot v^{H}(y_{11}) \right];$$

$$v^{C}(z_{2}) = \left[v^{B}(y_{5}) \cdot v^{6C}(y_{6}) \cdot v^{C}(y_{7}) \cdot v^{6C}(y_{8}) \cdot v^{C}(y_{9}) \cdot v^{C}(y_{10}) \cdot v^{6C}(y_{11}) \right] \vee \tag{4}$$
$$\vee \left[v^{B}(y_{5}) \cdot v^{6C}(y_{6}) \cdot v^{C}(y_{7}) \cdot v^{C}(y_{8}) \cdot v^{C}(y_{9}) \cdot v^{C}(y_{10}) \cdot v^{6C}(y_{11}) \right] \vee$$
$$\vee \left[v^{B}(y_{5}) \cdot v^{6C}(y_{6}) \cdot v^{C}(y_{7}) \cdot v^{C}(y_{8}) \cdot v^{C}(y_{9}) \cdot v^{H}(y_{10}) \cdot v^{H}(y_{11}) \right];$$

$$v^{6C}(z_{2}) = \left[v^{C}(y_{5}) \cdot v^{H}(y_{6}) \cdot v^{C}(y_{7}) \cdot v^{B}(y_{8}) \cdot v^{B}(y_{9}) \cdot v^{HC}(y_{10}) \cdot v^{H}(y_{11}) \right] \vee$$
$$\vee \left[v^{C}(y_{5}) \cdot v^{H}(y_{6}) \cdot v^{C}(y_{7}) \cdot v^{B}(y_{8}) \cdot v^{C}(y_{9}) \cdot v^{HC}(y_{10}) \cdot v^{H}(y_{11}) \right] \vee$$
$$\vee \left[v^{C}(y_{5}) \cdot v^{H}(y_{6}) \cdot v^{C}(y_{7}) \cdot v^{B}(y_{8}) \cdot v^{C}(y_{9}) \cdot v^{HC}(y_{10}) \cdot v^{H}(y_{11}) \right];$$

$$v^{B}(z_{2}) = \left[v^{C}(y_{5}) \cdot v^{H}(y_{6}) \cdot v^{B}(y_{7}) \cdot v^{B}(y_{8}) \cdot v^{C}(y_{9}) \cdot v^{HC}(y_{10}) \cdot v^{H}(y_{11}) \right] \vee$$
$$\vee \left[v^{HC}(y_{5}) \cdot v^{6C}(y_{6}) \cdot v^{B}(y_{7}) \cdot v^{B}(y_{8}) \cdot v^{C}(y_{9}) \cdot v^{HC}(y_{10}) \cdot v^{H}(y_{11}) \right] \vee$$
$$\vee \left[v^{C}(y_{5}) \cdot v^{H}(y_{6}) \cdot v^{B}(y_{7}) \cdot v^{B}(y_{8}) \cdot v^{C}(y_{9}) \cdot v^{HC}(y_{10}) \cdot v^{H}(y_{11}) \right].$$

In this context, weights of rules characterizing judgmental measures of expert certainty in the rule are not specified. The matter is under rough tuning their values are unit. In general, each input linguistic variable $y_1 - y_{12}$ has its own membership function to qualitative fuzzy terms (H, нC, C, вC, and B) to be used in equations 2 – 4. To simplify modeling, it is expedient to use one common form of membership function for each $y_1 - y_{12}$ variable. To do that standardization of intervals of linguistic variables to one general interval $\{0,4\}$ is performed with the help of standard ratios.

Selection of membership functions takes place within reliability; besides, they are accurately approximated by expert through the method of paired comparisons.

Fuzzy logic equations (2-4) together with membership functions of fuzzy terms are used as decision-making algorithm as for residual resource of RBT-type diamond crowns in

accordance with standard algorithm of intellectual technologies of identification (Meshcheriakov, 2007).

It should be noted that in fact, decision concerning differential analysis is a selection of one or several available alternatives. Any decision should be determined as a fuzzy set in a scope of alternatives found out as a result of intersection of specified goals and limitations. First, idea of decision as a fuzzy set in a scope of alternatives may be considered as artificial one. However, fuzzy decision may be considered as certain direction which fuzziness is the result of rough characterization of specified goals and limitations. In this context, the decision precision to specified goals is identified in each case through a formed value of adequate membership function. For example, for mode technological and technical parameters of DF within operating time sample of drilling process

$$y_1^* = 1000 \text{ kgf} \qquad y_2^* = 23.1 \ c^{-1}; \qquad y_3^* = 60 \text{ liters per minute} \qquad y_4^* = 0.99 \text{ meters per hour}$$

$$y_5^* = 0.071 \text{ mm per one revolution} \qquad y_6^* = 1100 \text{ m} \qquad y_7^* = 18 \text{ hours} \qquad y_8^* = 26.85 \text{ hours}$$

$$y_9^* = 10 \qquad y_{10}^* = 9.9 \text{ karat} \qquad y_{11}^* = 1.55 \text{ mm} \qquad y_{12}^* = 58.5 \text{ UAH}$$

there are found out values of standard membership functions within y_i^* points, $i = \overline{1,12}$ for each term in accordance with mode

technological and technical parameters of residual resource state of RBT-type of diamond crowns (Table 5).

Table 5. Values of standard membership functions within y_i^* points, $i = \overline{1,12}$ for each term in accordance with mode technological and technical parameters of residual resource state of RBT

	y_i^*	u_i^*	$v^H\left(y_j^*\right)$	$v^{HC}\left(y_j^*\right)$	$v^C\left(y_j^*\right)$	$v^{вC}\left(y_j^*\right)$	$v^B\left(y_j^*\right)$
1	1000	2	0,176	0,46	1,0	0,46	0,176
2	23,1	5,938	0,024	0,034	0,052	0,09	0,185
3	60	1,867	0,196	0,531	0,98	0,399	0,158
4	0,99	4,0	0,051	0,086	0,176	0,46	1,0
5	0,071	1,687	0,23	0,643	0,897	0,331	0,137
6	1100	0,4	0,842	0,703	0,25	0,112	0,062
7	18	2,286	0,14	0,34	0,913	0,625	0,225

	y_i^*	u_i^*	$v^H\left(y_j^*\right)$	$v^{HC}\left(y_j^*\right)$	$v^C\left(y_j^*\right)$	$v^{6C}\left(y_j^*\right)$	$v^B\left(y_j^*\right)$
8	26,85	3,566	0,063	0,115	0,258	0,727	0,819
9	10	4,0	0,051	0,086	0,176	0,46	1,0
10	9,9	1,247	0,354	0,933	0,601	0,217	0,101
11	1,55	0,0	1,0	0,46	0,176	0,086	0,051
12	58,5	1,07	0,427	0,994	0,496	0,186	0,09

Implementation of the system of fuzzy logic equations connecting standard membership functions of diagnosis and adequate input linguistic variables and qualitative terms as for mode parameters of technological area of DF is:

$$vHZ1 = \max\left[\min\left(v_{0,4}, v_{1,0}, v_{2,1}, v_{3,0}\right), \min\left(v_{0,2}, v_{1,0}, v_{2,1}, v_{3,0}\right), \min\left(v_{0,1}, v_{1,1}, v_{2,4}, v_{3,3}\right)\right];$$

$$vnCZ1 = \max\left[\min\left(v_{0,3}, v_{1,1}, v_{2,1}, v_{3,2}\right), \min\left(v_{0,3}, v_{1,2}, v_{2,1}, v_{3,2}\right), \min\left(v_{0,4}, v_{1,0}, v_{2,1}, v_{3,:}\right.\right.$$

$$vCZ1 = \max\left[\min\left(v_{0,3}, v_{1,0}, v_{2,0}, v_{3,4}\right), \min\left(v_{0,0}, v_{1,0}, v_{2,1}, v_{3,2}\right), \min\left(v_{0,3}, v_{1,0}, v_{2,0}\right.\right. \tag{5}$$

$$vbCZ1 = \max\left[\min\left(v_{0,2}, v_{1,4}, v_{2,2}, v_{3,4}\right), \min\left(v_{0,4}, v_{1,4}, v_{2,3}, v_{3,3}\right), \min\left(v_{0,3}, v_{1,4}, v_{2,2}, v_{3,4}\right)\right.$$

$$vBZ1 = \max\left[\min\left(v_{0,4}, v_{1,0}, v_{2,1}, v_{3,0}\right), \min\left(v_{0,2}, v_{1,0}, v_{2,1}, v_{3,0}\right), \min\left(v_{0,1}, v_{1,1}, v_{2,4}, v_{3,3}\right)\right]$$

As for the mode parameters of technical area of DF, the system of fuzzy logic equations is:

$$vHZ2 = \max\left[\min\left(v_{4,0}, v_{5,4}, v_{6,0}, v_{7,0}, v_{8,2}, v_{9,1}, v_{10,0}\right), \min\left(v_{4,0}, v_{5,0}, v_{6,1}, v_{7,0}, v_{8,4}, v_{9,1}, v_{10,0}\right),\right.$$
$$\left.\min\left(v_{4,0}, v_{5,2}, v_{6,2}, v_{7,1}, v_{8,4}, v_{9,1}, v_{10,0}\right)\right];$$

$$vnCZ2 = \max\left[\min\left(v_{4,2}, v_{5,2}, v_{6,2}, v_{7,1}, v_{8,2}, v_{9,0}, v_{10,0}\right), \min\left(v_{4,1}, v_{5,2}, v_{6,2}, v_{7,1}, v_{8,2}, v_{9,0}, v_{10,0}\right),\right.$$
$$\left.\min\left(v_{4,2}, v_{5,2}, v_{6,2}, v_{7,1}, v_{8,2}, v_{9,1}, v_{10,0}\right)\right];$$

$$vCZ2 = \max\left[\min\left(v_{4,4}, v_{5,2}, v_{6,2}, v_{7,2}, v_{8,2}, v_{9,2}, v_{10,3}\right), \min\left(v_{4,3}, v_{5,4}, v_{6,3}, v_{7,3}, v_{8,4}, v_{9,1}, v_{10,0}\right),\right. \tag{6}$$
$$\left.\min\left(v_{4,4}, v_{5,2}, v_{6,2}, v_{7,2}, v_{8,0}, v_{9,1}, v_{10,0}\right)\right];$$

$$vbCZ1 = \max\left[\min\left(v_{4,2}, v_{5,0}, v_{6,2}, v_{7,4}, v_{8,4}, v_{9,1}, v_{10,0}\right), \min\left(v_{4,1}, v_{5,2}, v_{6,4}, v_{7,4}, v_{8,2}, v_{9,1}, v_{10,0}\right),\right.$$
$$\left.\min\left(v_{4,2}, v_{5,0}, v_{6,2}, v_{7,4}, v_{8,2}, v_{9,1}, v_{10,0}\right)\right];$$

$$vBZ2 = \max\left[\min\left(v_{4,2},v_{5,0},v_{6,4},v_{7,4},v_{8,2},v_{9,1},v_{10,0}\right),\min\left(v_{4,2},v_{5,0},v_{6,4},v_{7,4},v_{8,2},v_{9,1},v_{10,0}\right),\right.$$
$$\left.\min\left(v_{4,1},v_{5,3},v_{6,4},v_{7,4},v_{8,2},v_{9,1},v_{10,0}\right)\right].$$

As for the diagnosis of residual resource of RBT of $x_1 - x_5$ diamond crowns of 01АЗД20К30 – 14А4Д50К50 types, the system of fuzzy logic equations is:

$$vX1X = \max\left[\min\left(vHZ1,vHZ2,v_{11,0}\right),\min\left(vHZ1,vnCZ2,v_{11,1}\right),\min\left(vnCZ1,vnCZ2,v_{11,0}\right)\right];$$

$$vX2X = \max\left[\min\left(vnCZ1,vnCZ2,v_{11,1}\right),\min\left(vCZ1,vnCZ2,v_{11,1}\right),\min\left(vnCZ1,vnCZ2,v_{11,2}\right)\right];$$

$$vX3X = \max\left[\min\left(vCZ1,vnCZ2,v_{11,2}\right),\min\left(vbCZ1,vbCZ2,v_{11,1}\right),\min\left(vbCZ1,vCZ2,v_{11,2}\right)\right]; \quad (7)$$

$$vX4X = \max\left[\min\left(vCZ1,vBZ2,v_{11,2}\right),\min\left(vbCZ1,vbCZ2,v_{11,4}\right),\min\left(vBZ1,vbCZ2,v_{11,3}\right)\right];$$

$$vX5X = \max\left[\min\left(vBZ1,vBZ2,v_{11,4}\right),\min\left(vbCZ1,vBZ2,v_{11,3}\right),\min\left(vBZ1,vBZ2,v_{11,3}\right)\right].$$

The greatest value of standard membership functions $vX1X, vX2X, vX3X, vX4X, vX5X$ in accordance with mode technological and technical parameters of resource state of RBT-type of diamond crowns is used to determine their operative diagnosis in pursuant to the set parameters $x_1 - x_5$.

Software to implement the algorithm is independent module (Meshcheriakov, 2007). It has been applied with the help of programming language Delphi; besides, it helps to compare verificatory diagnosis, and model one. Table 2 shows the results of comparative analysis of residual state of RBT of diamond crowns of 01АЗД20К30 – 14А4Д50К50 types according to such diameters as D=76 mm, D=93 mm, and D=112 mm on control sampling of 37 values.

For the research, simple average of absolute error while comparing model and verificatory diagnosis is $\bar{\varepsilon} = 0.1621621$ for diamond crowns of 01АЗД20К30-14А4Д50К50 D=76 mm and D=93 mm types. Thus, average error which identifies accuracy of resource state of RBT-type of diamond crowns is $\bar{\varepsilon}_x = 0.0266592$ (2.66%) for diamond crowns of 01АЗД20К30-14А4Д50К50 types (if D=76 mm and D=93 mm), and $\bar{\varepsilon}_x = 0.0202333$ (2.02%) for the same type of diamond crowns if D=112 mm.

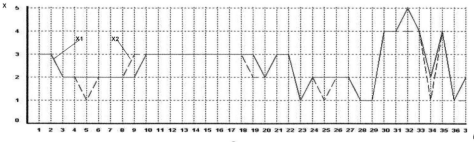

a

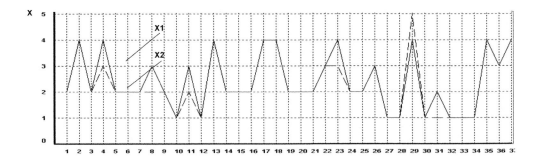

b

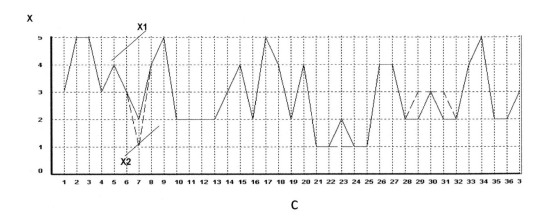

c

Figure 2. The comparison of verificatory ($X1$) and model ($X2$) diagnosis of residual resource of diamond crowns of 01А3Д20К30-14А4Д50К50 D=76 mm (a), D=93 mm (b), and D=112 mm (c) types on control sampling of mode and technological parameters.

CONCLUSIONS

Hence, basic principles of input-output variables linguisticity which estimate inputs and outputs of DF through qualitative terms while forming input-output dependence structures in the form of fuzzy knowledge base on collect of rules reflecting practices of mining experts and hierarchical pattern of knowledge bases give ability to overcome size restriction of opinion system through their classification and deduction tree derivation on the system of opinions and less knowledge interfolding. Together that stipulates confidence of early differential diagnosis goal setting. In addition, the use of fuzzy sets theoretical framework provides formalizing and applying of linguistic fuzziness of technological and technical states of basic functional structures of DF. By the same procedure, probability theory formalizes and applies statistical uncertainties for similar problems. Such a formulating of uncertainty state provides sufficient for mining practice accuracy application of intellectual technology of identification in automated systems of technological processes control.

REFERENCES

Meshcheriakov, L.I. 2006. *Analytical development of regulators of technological processes of drilling facilities through modified method of symmetry* (in Russian). Scientific bulletin NGAU, Issue 1: 85–91.

Dudlia, M.A., Karpenko, V.M. & Gryniak, O.A. 2005. *Drilling processes automation* (in Russian). Dnipropetrovs'k: NMU: 207.

Meshcheriakov, L.I. & Klochkov, Yu.V. 2007. *Software for the systems of intellectual support for drilling facilities control* (in Russian). Mining electrical engineering and automation: 68–75.

Energy Efficiency Improvement of Geotechnical Systems – Pivnyak, Beshta & Alekseyev (eds)
© 2013 Taylor & Francis Group, London, ISBN 978-1-138-00126-8

Two-propulsion travelling mechanisms of shearers for thin beds

N. Stadnik & V. Kondrakhin
Donetsk National Technical University, Donets'k, Ukraine

L. Tokar
State Higher Educational Institution "National Mining University", Dnipropetrovs'k, Ukraine

ABSTRACT: The paper deals with urgent for coal industry problem of designing of shearers for thin beds. The shearers should be equipped with travelling mechanisms with two frequency-controlled drives, one frequency converter, and rigid traction unit. The paper deduces from the experiments and theoretically explains a phenomenon of originating of out-of-phase load variations within drives of such a type. The paper gives recommendations on decrease in affect of load variations on shearer efficiency. The paper proposes a version of correcting a proportional control law in frequency-controlled drive, providing extension of a range of service speeds of the shearer as well as required moving forces.

1 INTRODUCTION

Experience of new shearers designing and applying assures that they should provide working travel of machines at a speed of up to 10-15 meters per minute; moving force should be up to 300-400 kN within faces which length is up to 300-400 meters.

Under the conditions of reduced clearances of machines, such high rates are possible only owing to application of electrical travelling mechanisms. Such integral electrically driven travelling mechanisms with frequency-controlled asynchronous drive and rigid traction unit as УКД300, УКД400, and КДК500 (Ukraine), SL300 and EL600 (Germany), 7LS1and 7LS3 (the USA) etc. are widely used in modern mine-building.

Together with other factors, applications of the above-mentioned electric drive of travelling mechanisms identify transformation of modern shearers to mechatronic machine (Stadnik, N.I., Sergeiev, A.V. & Kondrakhin, V.P. 2007). Advanced travelling mechanism of a shearer with frequency-controlled drive is a mechatronic assembly covering all components being typical for such devices. Components of different physical nature incorporated in one assembly originate a problem of their interacting which provides required traction-speed and resource characteristics. Availability of two interacting frequency-controlled electric drives is the travelling mechanisms feature.

Practice of such shearers designing and exploitation shows (Kosariev, A.V., Stadnik, N.I., Sergeiev, A.V., Kondrakhin, V.P., Lysenko, N.M. & Kosariev, V.V. 2007; Kondrakhin, V.P., Lysenko, N.M., Kosariev, A.V., Kosariev, V.V. & Stadnik, N.I. 2006) that in a number of cases uncoordinated performance of movers takes place. The fact prevents achievement of planned tractional factors as well as the machine life.

As it is seen hereafter, the incardination depends on out-of-phase load variations in drives.

Besides, relative to safety requirements, axial drive is dual-motor with a single frequency converter for downsizing. That is why it is practically impossible to apply closed-loop controls (both scalar and vector).

In this connection, applied open-loop control mode $U_1 / f_1 = const$ (proportional control) should be corrected.

Problems of interacting and coordinated operations of electric drives of shearers travelling mechanisms are considered to be the least studied, and their analysis determines the paper content notably.

Mechatronics-based approach to formation of a law of frequency-controlled drive of shearers providing required control range, traction and speed characteristics, and overload capability is introduced.

2 FORMULATING THE PROBLEM

Loads in frequency-controlled drive of two-propulsion travelling mechanisms of shearers with rigid tractional unit are of dynamic nature. Firstly, the nature depends on inequality of resistive forces of machine movement (forces on operative devices and friction in bearing parts). Secondly, it depends on sprocket-rod action character.

Accordingly, one may provisionally distinguish two components of loads in drives.

Oscillations matching first components take place within two sprocket drives in phase.

Availability of in-phase and out-of-phase oscillations is typical for second component. Their period is equal to one sprocket tooth overlapping.

The oscillations amplitudes depend on a phase shift value between start of engagement teeth of driving sprockets with pin rode, mean shifting speed, and rated slip of motor sets.

Low-frequency oscillations which period is equal to a period of a shearer move over a distance similar to a length of a rod section are also typical for second component (Kosariev, A.V., Stadnik, N.I., Sergeiev, A.V., Kondrakhin, V.P., Lysenko, N.M. & Kosariev, V.V. 2007; Kondrakhin, V.P., Lysenko, N.M., Kosariev, A.V., Kosariev, V.V. & Stadnik, N.I. 2006).

Sustained moment value plays important part for drives having dynamic loading condition while identifying their parameters, initial data for stress analysis, and function. The value is maximum value of average level of motor set torque when it can operate steadily, without stalling under static conditions.

Paper (Starikov, B.Ya., Azarkh, V.L. & Rabinovich, Z.M. 1981) proposes expressions to identify sustained moment of actuator of shearer taking into consideration randomness of acting loads.

Paper (Starikov, B.Ya., Azarkh, V.L. & Rabinovich, Z.M. 1981) gets similar expressions for dual-motor drive of operative devices of a shearer taking into account engine scatter, and its dynamic properties.

However, to assess sustained moment of dual-motor frequency-controlled drive of travelling mechanism, the results may be applied only after further research.

That depends on peculiarities of load oscillation formation in travelling mechanisms, and their low-frequency nature.

Besides, a problem of speed frequency regulation range plays important part for travelling mechanism drive. The problem is directly connected with a value of sustained moment of a drive.

The paper formulates and solves the problem of assessment of sustained moment, and a range of frequency control of dual-motor drives of shearer travelling mechanism to make grounded choice of their parameters.

3 RESULTS OF INVESTIGATION

Frequency-controlled drive of shearers move (Stadnik, N.I., Sergeiev, A.V. & Kondrakhin, V.P. 2007) covers two identical axial drive units $УПП_i$ (Figure 1), supplied by electric energy transducer (EET). They consist of electromechanical transducer $ПЭВ_i$ (asynchronous motor), mechanical transducer $ПВВ_i$ (speed reducer), and mechanical transducer $ПВП_i$ (three-element mover "sprocket – pin sprocket – pin rod").

Input control function $\varepsilon_{ex}(V_n)$ which value identifies set speed V_n of shearer supply is transformed by EET electronic component into electric signal ε_1 with running voltage U_1 and frequency f_1 values.

Output function $\mu_{вых}$ quantitatively determined by means of travelling force Y_n characterizes mechanical step of a shearer.

Modules $ПЭВ_i$ represent electromechanical components. Electrical energy ε_I with U_1 and f_1 converts into rotational movement – interface ω_{i1} characterized by moment $M_{Дi}$ and rotating speed $n_{Дi}$.

Modules $ПВВ_i$ and $ПВП_i$ represent mechanical component. Modules $ПВВ_i$ convert interface ω_{i1} into interface ω_{i2} - rotational movement into rotational movement with other parameters. Modules $ПВП_i$ convert mechanical rotational movement into mechanical step – function μ_i. Tractive force is realized while adding forces developed by $УПП_i$.

Speed-control range, required tractive efforts, and overload capability to be important owing to dramatic load dynamics are basic parameters of axial drive.

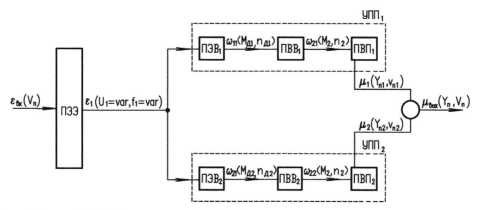

Figure 1. Structural circuit of frequency-controlled axial drive

Travelling mechanisms of such shearers for thin beds as УКД300 and УКД400 are considered as test objects.

Due to serious space limitations, problems of choosing rational parameters of a drive are of great importance for such machines.

Basic parameters of travelling mechanisms of УКД300 and УКД400 shearers are in Table 1.

Studies of results of load measurement in dual-motor axial drives of УКД300 and УКД400 shearers show that such loads as motor currents and moments have random oscillating components.

Highest possible oscillating frequency increases proportionally to mean travel speed not exceeding 1.6 Hz.

Table 1. Parameters of travelling mechanisms

Parameter	УКД300	УКД400
Mover	Three-element	Two-element
Rack arrangement	From a goaf side of conveyer	From a face side of conveyer
Distance between sprocket axes, mm	2160	882
"Sprocket-rack" gear pitch, mm	100	126
Axial drive power, kW	2x30	2x30
Maximum operating speed, meters per minute	8,5	8
Maximum creep speed, meters per minute	12	12
Maximum tractive force, kN	300	300

Figure 2 shows oscillograph of torques of left and right electric motors of УКД300 shearer axial drive as a result of rig tests.

It should be noted that under rig conditions, forces of cutting and feeding on operating devices were simulated by approximately constant, static resistance.

As Figure 2 shows, important out-of-phase load variations are formed in drives of travelling mechanism of УКД300 shearer.

Period of high-frequency components is equal to a period of sprocket tooth-rack gearing. As for low-frequency, it is equal to a period of a shearer move over a distance similar to length of one spout (1.5 m).

Papers (Kosariev, A.V., Stadnik, N.I., Sergeiev, A.V., Kondrakhin, V.P., Lysenko, N.M. & Kosariev, V.V. 2007; Kondrakhin, V.P., Lysenko, N.M., Kosariev, A.V., Kosariev, V.V. & Stadnik, N.I. 2006) interpret the oscillations nature.

As distance between sprocket axes of УКД300 shearer is reasonably large (2.160 m), and there are gaps between rack sections, sprocket teeth gears rack out of synchronization. That can explain availability of out-of-synchronization load variations with a period equal to a period of tooth gear (about 1.5 s) (in Figure 2).

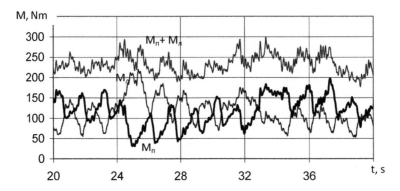

Figure 2. Oscillogram of torques of left Мл and right M_n of electric motors, and total moment of УКД300 shearer axial drive, when its speed is 4.2 meters per minute

Figure 3 demonstrates oscillogram of current of left and right electric motors of УКД400 shearer axial drive obtained in the process of measurement in a mine.

The oscillogram is recorded while coal mining at a speed of 5 meters per minute if amount of inclination is 0°.

A curve of both motors total current is also shown here.

The singularity of formation of a drive of УКД400 shearer travelling mechanism is that during some period (more specifically, 5/12 of total operating time conforming to sprocket gear with five teeth of twelve placed on one section of rack) both sprockets are geared with one rack teeth.

During other periods, sprockets are geared with neighbouring racks teeth.

If sprockets contact with different racks teeth, starting points of teeth gearings are mistimed due to gaps between racks. Phase displacement between start of teeth gearing originates resulting in out-of-phase load variations in drives with considerable amplitude.

Above-mentioned modes of loads forming correspond to 0-12-s section in Figure 4.

As distance between sprockets axes is taken as that multiple as a rack teeth distance (ratio 7), under similar wear degree, both sprockets teeth gear with one rack teeth is synchronous.

In this case, load variations in both drives are in-phase having comparatively small amplitude (section 12-20 s in Figure 2).

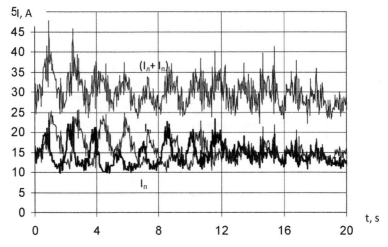

Figure 3. Oscillogram of left and right electric motors current of УКД400 shearer axial drive when V_n=5meters per minute

206

As both drives of travelling mechanism work in common tractional unit, then electric motors stalling (on-load stop) will take place simultaneously despite difference in momentary values of both engines in steady-state.

Figure 4 shows a fragment of oscillogram of current while stalling of electric motors of travelling mechanism of УКД400 shearer.

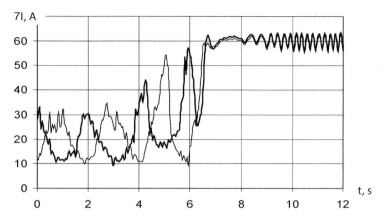

Figure 4. Oscillogram of current while stalling of electric motors of travelling mechanism of УКД400 shearer

As Figure 4 demonstrates, current oscillations of electric motors take place in opposition around average level close to rated current 24.5 A until their values approach critical value.

After that, increase in both motors current occurs more or less synchronously.

It follows that to assess sustained moment of dual-motor drive of travelling mechanism one should take into consideration irregularity of total load moment.

Figures 2 and 3 mean that evolution of total load of axial drive of shearer should be considered as random.

To assess sustained moment as a parameter characterizing spread of values of random process as for average one, paper (Starikov, B.Ya., Azarkh, V.L. & Rabinovich, Z.M. 1981) proposes to take into account load variation factor.

Oscillogram sections with roughly constant level of mean load and constant travelling speed were taken for statistical analysis.

Minimum realization length was taken as such to make a distance by a shearer close to length of one conveyer spout (1.5 m).

Chosen realization length of random process helps to take into account peculiarities of loads forming from the viewpoint of their amplitudes and variation factor.

Results of the statistical analysis show that on a first approximation, within 0.6 to 1.6 current range of rated value, current and a moment in

operative current range have direct proportion, and proportion factor is about 12 Nm/A.

Variation factor (VF) of every motor of УКД300 shearer travelling mechanism increases with speed increase varying within 0.200 (1.7 meters per minute (10 Hz) to 0.365 (4.2 meters per minute (25Hz).

With it, VF of total moment stays more or less constant being 0.1; this points to the fact that on the whole load on a drive is constant.

Statistical analysis shows that VF of loads of every motor of УКД300 shearer travelling mechanism is 1.08 to 2.07 times greater to compare with УКД400 shearer.

The greatest VF drop of УКД400 shearer load is within the most typical rate of mining being 4 to 5 meters per minute (1.7 to 2.07 times).

It should be noted that real effect of VF drop of УКД400 shearer is higher to compare with specified as when loads of УКД300 shearer were measured, dynamic components of cutting and loading forces were not available.

That confirms effective strength of engineering decisions made to drop out-of-phase load variations of УКД400 shearer travelling mechanism (centers between sprockets are taken as multiple of gear pitch and as low as practicable, traction unit is located from a face, etc.).

At that very time variation factors of total loads of УКД400 shearer travelling mechanism are 1.1 to 1.3 times greater. It depends on load difference resulting from feeding forces on operating devices while coal cutting and loading.

As it is above-mentioned, jig tests of УКД300 shearer showed zero difference.

Available experimental data can not make it possible to determine a dependence of variation factor of total load of shearer travelling mechanism v_c on average displacement speed (on feeding frequency).

Hence, to assess sustained moment with stability margin, one may take maximum VF value being v_c=0.13 for УКД400 shearer.

The value corresponds to meaningful operational conditions of a shearer. For this reason, it may be assumed as basic one while calculating sustained moment of travelling mechanism drive.

To determine sustained moment, the following assumptions are taken:

- Due to low-frequency (up to 1.6 Hz) character of oscillations, dynamic properties of electric motor may be neglected as natural frequency of asynchronous electric motor is about 11 Hz slightly depending on power frequency.

- Possible scatter of nominal parameters of engines decreasing in the process of heat is ignored.

- With rather high probability, random values of total load of a drive are not greater than M_c $(1+3v_c)$, where M_c is estimation of expectation (average level).

Hence, sustained moment of dual-motor drive can be determined by

$$M_y(\alpha) = \frac{M_\kappa(\alpha)}{1 + 3v_c} \qquad (1)$$

where $M_\kappa(\alpha) = 2 \cdot M_{\kappa\partial}(\alpha)$ is critical moment of dual-motor drive; $M_{\kappa\partial}(\alpha)$ is critical moment of single electric motor; $\alpha = f_1 / f_{1HOM}$ is current proportional frequency of stator; f_1 – voltage frequency of stator; and $f_{1HOM} = 50$ Hz.

Paper (Chilikin, M.G. & Sandler, A.S., 1981) proposes a formula to determine values of critical moment of asynchronous electric motor at any artificial characteristic:

$$M_{\kappa\partial}(\alpha) = \frac{m_1 \cdot U_{1H\phi}^2 \cdot \gamma^2}{2 \cdot \omega_{1HOM}} \cdot \frac{1}{r_1\alpha + \sqrt{(b^2 + c^2\alpha^2)(d^2 + e^2\alpha^2)}} \qquad (2)$$

where $U_{1H\phi}$ is actual value of coil voltage of engine if $f_{1HOM} = 50$ Hz; m_1=3 is number of stator phases; ω_{1HOM} is a normal angular of frequency of revolution; $\gamma = U_1/U_{1H\phi}$ is a stator relative voltage; U_1 is a stator voltage; $\tau_1 = x_1/x_0$ is dissipation coefficient of a stator; $\tau_2 = x'_2/x_0$ is

dissipation coefficient of a rotor; $\tau = \tau_1 + \tau_2 + \tau_1 \tau_2$ is gross dissipation coefficient; $b = r_1 (1 + \tau_2)$; $c = x_0\tau$; $d = r_1/x_0$; $e = (1 + \tau_1)$ are coefficients depending on equivalent circuit parameters; r_1, x_1 are coil resistance and inductive resistance of a stator; x'_2 is reduced inductive resistance of a rotor; and x_0 is reactive impedance of magnification circuit.

For ЭКВ4-30-6-02 electric motor parameters are: $U_{1H\phi} = 548$ V, $\omega_{1HOM}= 103$ c^{-1}, $r_1= 1,09$ Ohm, $x_1= 2,35$ Ohm, $r_2= 0,724$ Ohm, $x_2= 2,50$ Ohm, and $x_0= 53$ Ohm.

As a rule, open-loop control system by law of frequency regulation U_1/f_1=const (or $\gamma = \alpha$) is taken up in travelling mechanisms of shearers while controlling down of nominal frequency ($\alpha < 1$).

If $\alpha > 1$, then control law U_1 = const is taken up. Efficient use of more complex feedback-control systems (both scalar, and vector) is made much difficult by the fact that travelling mechanism drive is dual-motor. With it, one frequency rectifier powers both engines. Such a design of drive depends on stringent requirement for its overall dimensions (especially height).

At the voluntary moment, both motor load and rotational frequency may differ greatly (Figure 2 and 3); that makes inefficient closed-loop control on current or on rotational frequency of either motor. If two frequency converters are applied in a roadway, then individual cable for each motor is required. It can hardly be carried out in practice. Due to severe space limitations, mounting of two on-board changers is possible only for high-coal shearers. Hence, in the majority of cases, it is expedient to apply open-circuit systems of variable-frequency control for travelling mechanisms of shearers operating in thin beds.

To plot $M_{\kappa\partial}(\alpha)$ characteristic for two-region control in expression (2), one should assume $\gamma = \alpha$ if $\alpha < 1$, and $\gamma = 1$ if $\alpha > 1$. Figure 5 shows $M_\kappa(\alpha)$ и $M_y(\alpha)$ dependence diagrams for parameters of a drive with two ЭКВ4-30-6-02 motors if v_c=0,13.

Marks "□" in Figure 5 mean the points obtained as a result of experimental research of travelling mechanism of УКД300 shearer on a loading jig under steady loading. The points correspond closely to a $M_\kappa(\alpha)$ curve. That confirms ability to apply expression (2) for assessment of critical moment of electric motors of travelling mechanisms of shearers.

Figure 5 also demonstrates torque rating M_n of dual-motor drive (dot line) which is identified with the help of warming-up allowance of engine under continuous service if supply frequency is nominal (that is α = 1). In this context, it is assumed on a first approximation that a value of allowed warming-up moment within considered range does not depend on supply frequency (or on α). For motors with independent cooling (water cooling in this instance) when control is performed down of nominal frequency, the assumption is also practically assured. While controlling up of nominal frequency, extra research is expedient to check the assumption.

Intersections of curve $M_y(α)$ and line $M_н$ determine edges of control range $α_1$ and $α_2$ in which sustained moment of a drive is greater than rated one.

In the case under consideration, $α_1$=0.28 and $α_2$=1.28; that is torque rating of a motor may be implemented under 14 to 64 Hz supply frequency. Obtained limit values α should correspond to minimum and maximum working speed of travelling mechanism.

Obtained curves show that under control mode $γ = α$ critical moment value decreases when frequency drops. To keep it constant within control range, less voltage is required at small frequencies to compare with frequency drop.

Consider ability of lower frequency drop (14 Hz), and at the expense of that to extend range of speed control. To analyze abilities of a drive, and to adopt dependencies of transformer voltage on frequency, apply equivalent circuit shown in

Figure 6. Method (Chilikin, M.G. & Sandler, A.S. 1981) is used for motor moment estimation.

Following symbols are applied while calculating.

Absolute slip parameter $β$, or relative frequency of a rotor – absolute slip $Δω$ ratio to synchronous angular rate $ω_{1н}$ at nominal frequency

$$β = Δω / ω_{1н} = f_2 / f_{1н},\tag{3}$$

where f_2 is current frequency of a rotor.

Parameter $β$ is used instead of s slip and correlates with it in such a way

$$s = Δω / ω_1 = β / α.\tag{4}$$

Dissipation coefficient for stator and rotor is respectively

$$τ_1 = x_1 / x_m \text{ and } τ_2 = x_2' / x_m.\tag{5}$$

Total dissipation coefficient is

$$τ_1 = τ_1 + τ_2 + τ_1τ_2.\tag{6}$$

Set of equations (7) identifying performance of frequency-regulated drive of supply is:

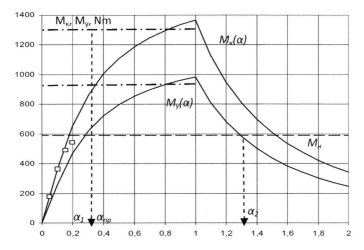

Figure 5. Critical moment and sustained moment of a drive of travelling mechanism with two electric motors ЭКВ4-30-6-02

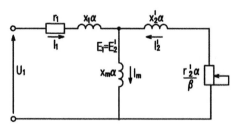

Figure 6. Equivalent circuit of asynchronous motor under variable-frequency control

$$
\left\{
\begin{array}{l}
M_{Дi} = \dfrac{m_1 U_1^2}{\omega_{1н}} \cdot \dfrac{r_2' \beta}{(b^2 + c^2 \alpha^2)\beta^2 + 2 r_1 r_2' \alpha\beta + (d^2 + e^2 \alpha^2) r_2'^2} \\[4mm]
M_{Д.кi} = \dfrac{m_1 U_1^2}{2\omega_{1н}} \cdot \dfrac{1}{r_1\alpha + \sqrt{(b^2 + c^2 \alpha^2)(d^2 + e^2 \alpha^2)}} \\[4mm]
n_{Д_i} \equiv \omega_{Дi} \equiv f_1 \\[2mm]
Y_{ni} = \dfrac{M_{Дi}}{r_{ci}} u \eta_n \\[3mm]
n_{Дi} = \dfrac{V_n}{r_{кi}} u \\[3mm]
Y_n = \sum Y_{ni}
\end{array}
\right\}
\qquad i = 1,2 \qquad (7)
$$

where m_1 is the number of stator phases; $M_{Дкi}$ is maximum torque of a motor; $b = r_1(1 + \tau_2)$; $c = x_m \tau$; $d = r_1 / x_m$ are coefficients depending on equivalent circuit parameters; x_m is reactive impedance of magnification circuit; $\omega_{Д_i}$, $n_{Д_i}$ are angular rate and motor speed; u is reduction ratio; r_{ci}, $r_{кi}$ are reduced power gearing radius and kinematic gearing radius (Sandler, A.S. & Sarbatov, R.S., 1974); Y_{ni} is moving force of each $УПП_i$; and Y_n is total moving force.

According to (Kondrakhin, V.P., Lysenko, N.M., Kosariev, A.V., Kosariev, V.V. & Stadnik, N.I. 2006), static moving force $Y_{n.c}$ of axial drive is determined in such a way:

$$
Y_{n.c} = k_f \left(G(\sin\alpha \pm f'' \cos\alpha) + \sum_{i=1}^{N_u} Y_{u.i} \right), \qquad (8)
$$

where k_f is a coefficient taking into account added resistance of a shearer move (is assumed as equal to 1.4); α is a pitch angle of a face line; f'' is a shearer friction coefficient (on a sill – 0.35; on transfer tracks – 0.21); G is a shearer

mass; $\sum_{i=1}^{N_u} Y_{u.i}$ is total feed force on operating device; and N_u is the number of operating devices.

In this context, as it is shown hereinafter, control mode is formulated on the basis of practicable traction performance of a drive allowing decrease in a motor overload capability being less than rated one, nevertheless providing sustained performance of a drive.

In terms of idealized traction and speed characteristic of a drive, supply drive power P_n (S1 mode) is identified by:

$$
P_n = \frac{Y_{n.c.\max} V_{n.\max}}{\eta_n}, \qquad (9)
$$

where $V_{n.\max}$ maximum axis velocity; $Y_{n.c.\max}$ is static moving force adequate to $V_{n.\max}$; the coefficient is calculated on (1); and η_n is a drive transmission efficiency (assumed as 0.75 in calculations).

According to (9), $P_n = 57.5$ kW in terms of $Y_{n.c.\max} = 22$ t and $V_{n.\max} = 12$ meters per minute ($-\alpha = 10°$ for meaningful operation conditions, and

210

cuttability is 240 N/mm); hence, power of singular machine axis drive motor is assumed as equal to 30 kW (57.5/2). With it, specified motor should have adequate overload capability (2.5-2.7). On the basis of the requirements, ЭКВ4-30-6-02 motor has been designed.

Now it is required to make a choice of a drive ration depending on a number of inter-related issues (which concern the design including parameters of sprocket, rake etc.) which the paper does not consider. Ultimately, noted ratio identifies axis velocity $V_{n.н}$ being adequate to nominal frequency of converter $f_{1н}$. This paper assumes it as $V_{n.н}$ =8 meters per minute. It should be noted that designing must determine the parameter with the help of iteration method each time forming adequate law of statutory variable-frequency control to provide required overload capacity of a drive over a range of speed regulation.

Noted law is determined analytically by means of a motor torque moment $M_{Д.кi}$ values calculation, adequate values of moving force $Y_{n.кi}$, implemented overload capacity λ_p (in relation to static moving force $Y_{n.c}$ for the axis velocity):

$$\lambda_p = \frac{Y_{n.кi}}{Y_{n.c}}. \qquad (10)$$

In this case, required moving forces should not be greater than rated values of a motor warming-up calculated on approach (Chilikin, M.G. & Sandler, A.S. 1981). Necessary overload capacity can't be under 1.5. That is identified including a value of sustained moment $M_{n.ycm}$ (1) of machine axis drive motor (maximum possible mean value of drive torque by the motor specified at established loading dynamism), and adequate moving force $Y_{n.ycm}$ calculated on the assumption of meaningful value a motor coefficient of variation $v_{дв}$.

Formation of variable-frequency control law is performed by means of iteration method through assigning a number of values of frequency and voltage (in increment equal to 0.1 of value $U_{1н} / f_{1н}$, obtained experimentally). The law for sharer УКД400 axial drive is in Figure 3; voltage values are greater than those for proportional control law ($U_1 / f_1 = const$) for 2.5 to 50 Hz variations of stator frequency if rated current is kept. Dependences diagram $\gamma = f(\alpha)$ for considered control laws are in Figure 7.

Mechanical characteristics of axial drive developed for two above-mentioned voltage control laws of frequency are shown in Figure 8.

Values λ (loading capacity relative to torque rating) indicate the efficiency of proportional control law "correction" for natural and simulated characteristics equal to 2.1; 1.5; 0.911 and 2.6; 2.4; and 1.7 for such frequency values as 30, 15 and 7.5 Hz; in this regard value of the index is 2.6 according to the motor specifications.

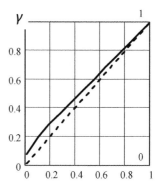

Figure 7. – Voltage dependences on frequency at the output of converter (dash line is for proportional control law $U_1 / f_1 = const$, and "corrected" is for control mode providing increase in overload capability)

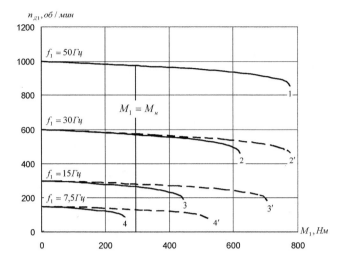

Figure 8. Mechanical characteristics of ЭКВ4-30-6-02 motor at frequency regulation: 1, 2, 3, and 4 are for control mode $U_1 / f_1 = const$; and 2', 3', and 4' are for control mode providing increase in the motor loading capacity

Figure 9 showing traction and speed characteristics of УКД400 shearer (α =10° for average operating conditions) demonstrates that maximum feeding speed is: 10 to 14.5 meters per minute in recovery mode depending on cuttability, and no less than 15 meters per minute in haulage mode if moving force is within the essential reserves. In this context, implemented loading capability λ_p is more than 1.5 per unit when cuttability is 360 N/m. The data belong to a drive ratio when velocity is 8 meters per minute at 50 Hz.

In this context, torque moment of the motor may be implemented within 7.5 to 64 Hz with corresponding change of shearer displacement speed within the regulation range.

Hence, the proposed law voltage control depending upon frequency provides required moving forces, axis velocity changes, and loading capability.

4 CONCLUSIONS

To assess sustained moment of dual-motor frequency-controlled drive of a shearer travelling mechanism, one should use values of variation coefficient of two drives total load. Variation coefficient of total load during recovery may be assumed as equal to 0.116…0.130 resulting from a mine measurement. The value of sustained moment of a drive travelling mechanism depends on supply frequency and control law type. While applying U_1/f_1 = const law, boundary values of per-unit frequency at which sustained drive moment with two ЭКВ4-30-6-02 motors is greater than its torque rating are equal to α_1=0,28 and α_2=1,28; that is torque rating of a motor may be implemented at 14 to 64 Hz. While applying "corrected" control mode, stability of a sustained drive moment is provided if control is performed down of nominal frequency to 7.5 Hz.

212

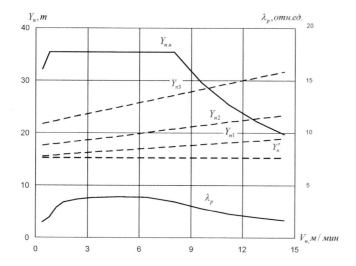

Figure 9. Dependences of moving forces on axis velocity in haulage mode Y_n', in recovery mode Y_{n1}, Y_{n2}, Y_{n3} (when cuttability is 120, 240 and 360 N/mm), allowable power on a motor warming-up $Y_{n.u}$, and implemented loading capability λ_p (if cuttability is 360 N/mm).

The paper approach to correction a law of voltage proportional control from frequency for a shearer axial drive based on mechatronics ideas makes it possible to formulate iteratively voltage dependence at the output of converter on frequency on the assumption of required control range and moving forces. The results of static mechanical characteristics calculation and implemented loading capability of electric drive are taken into consideration.

In future it is required to continue research into dynamic and thermal conditions of axial drive.

REFERENCES

Stadnik, N.I., Sergeiev, A.V. & Kondrakhin, V.P. 2007. *Mechatronics in coal machine-building* (in Russian). Mining equipment and electromechanics, Issue.4: 20-29.

Kosariev, A.V., Stadnik, N.I., Sergeiev, A.V., Kondrakhin, V.P., Lysenko, N.M. & Kosariev, V.V. 2007. *Comprehensive research of traveling mechanism of УКД300 shearer* (in Russian). Mining equipment and electromechanics, Issue.3: 2-6.

Kondrakhin, V.P., Lysenko, N.M., Kosariev, A.V., Kosariev, V.V. & Stadnik, N.I. 2006. *Simulation of loads in two-propulsion traveling mechanism of shearer with frequency-regulated drive* (in Russian). Scientific works of DonNTU, Vol.113: 139-145.

Starikov, B.Ya., Azarkh, V.L. & Rabinovich, Z.M. 1981. *Asynchronous drive of shearers* (in Russian). Moscow: Nedra: 288.

Chilikin, M.G. & Sandler, A.S. 1981. *Orientation course of electric drive: Textbook for Institutions of Higher Education* (in Russian). Moscow: Energoizdat: 576.

Sandler, A.S. & Sarbatov, R.S. 1974. *Automatic variable-frequency control of asynchronous motors* (in Russian). Moscow: Energia: 328.

Energy Efficiency Improvement of Geotechnical Systems – Pivnyak, Beshta & Alekseyev (eds)
© 2013 Taylor & Francis Group, London, ISBN 978-1-138-00126-8

Collective behaviour of automatic machines and the problem of resource allocation with limitation of "all or nothing" type

A. Zaslavski, P. Ogeyenko & L. Tokar
State Higher Educational Institution "National Mining University", Dnipropetrovs'k, Ukraine

ABSTRACT: A complex of problems of collective behaviour of automatic machines when a quantity of allocated resource is less than participants require is considered. Self-organization principle of automatic machines collective is studied. The analysis of centralized model of collective behaviour of automatic machines where each party follows "all or nothing" principle is given. As far as necessitated rejection of resource is concerned, damage limitation exercise has been worked out.

1 INTRODUCTION

Classic research of collective-behaviour models in the process of allocation of limited resource (Varshavski 1974) and (Burkov 1977) pays major attention to such problems in which participants of allocation may use any amount of resource within given limitations. However, there are many problems including game-theory ones (Petrosian et al. 2012) where the principle is not applicable and where the participants of allocation have to follow "all or nothing" principle. The paper considers the approach to solution of such problems by means of organizing the desired behaviour of automatic machines collective.

2 SETTING OF THE PROBLEM

Let's have a certain quantity of m-objects – resource consumers. During Δt time each ith object uses resource in $w_i(\Delta t) \geq 0$ amounts. It is assumed in this context that the objects can not be regulator components. It means that they have only two abilities. Either they operate using the resource within specified quantities, or they cease operations (they are off for Δt time). If during Δt time a resource owner can provide all

$$W = \sum_{i=1}^{m} w_i(\Delta t)$$ objects with the resource, then all

of them use it obtaining equal to $\psi_i = \psi(w_i)$ effect; otherwise, if $W < \sum_{i=1}^{m} w_i(\Delta t)$, then

disconnection problem arises when it is required to switch off a number of objects for the limitation period Δt. In this connection, switched-off objects bear $f_i = \psi_i - \lambda w_i \geq 0$ losses, where λ is the resource price. It is required to select a group of disconnecting objects $\alpha_1, \alpha_2, ... \alpha_i, ... \alpha_m, \alpha_i \in \{0,1\}$ in such a way to minimize damage caused by resource rejection

$$F = \sum_{i=1}^{m} (1 - \alpha_i) f_i \qquad (1)$$

And total resource demand should be within fixed limit

$$\sum_{i=1}^{m} \alpha_i w_i(\Delta t) \leq W \qquad (2)$$

Here α_i is Boolean variable which takes values 1 or 0 depending upon the fact whether i^{th} object obtains required resource amount or not. If it is admissible to consider $w_i(\Delta t)$, $\psi_i(\Delta t)$ and W values as constant at each disconnection, then the problem can be solved once for all occasions. However, in actual practice, parameters on which damage depends, experience random changes. Sometimes the changes may be essential, and well-defined disconnection group may be off-optimum. In this case it is possible to organize overall control for all objects. To do that, it is required to transmit to Control Centre data forecasted for disconnection period. The data show whether resource is essential for $w_i(\Delta t)$ objects or not. When the Centre receives the data, it makes decision on a group of users to be disconnected. Such an alternative is acceptable if the Centre is the resource seller which is not directly interested in its use efficiency. If the Centre stands to benefit not from the resource offer as it is but from the results of its use, then to make decision it also needs information concerning expected losses $f_i(\Delta t)$ by reason of disconnection. The process of collecting and processing large amount of information, while estimating predicted losses is the most complex part of the problem. That is why the major part of the system power controlling centralized allocation of the resource is used for information collecting and processing. The information source is located at the periphery. Even if the Centre owns the resource, it is interesting to have ability of self-organization of automatic machines collective which makes independent decisions concerning its participation in allocation of the resource.

But if consumers are collective owner of the resource, then the problem of its allocation by means of self-disconnection becomes urgent.

Any energy carriers, material resources, financial resources, and time may be considered as allocated resource within the scope of such problems. When information is transferred by means of packages (blocks) during a certain period of time which duration due to prevailing conditions is less than total period of informational packages, then the approach may be useful for automatic formation within limited period of optimum message blocks which comprise subsets of a given set of information packages.

3 SELF-ORGANIZATION PRINCIPLE OF AUTOMATIC MACHINES COLLECTIVE

Figure 1 shows architecture of a system controlling resource allocation.

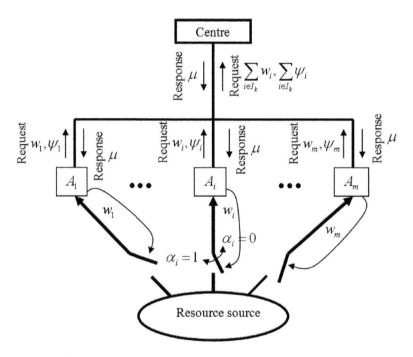

Figure 1. Architecture of a system of resource collective allocation

The system covers resource source which connection depends on A_1, A_2,... A_i,...A_m automatic machines. Each ith automatic machine can connect adequate consumer to ($\alpha_i = 1$) resource, or cease the process ($\alpha_i = 0$). To solve the problem, automatic machines inquire Centre. The requests include required quantity of resource w_i and desired efficiency of its use ψ_i. The Centre can not differentiate the requests. To control the system, it is sufficient to know at each

k^{th} preparation step of final allocation, a value of total inquired resource $\sum_{i \in I_k} w_i$, and total effect of its use $\sum_{i \in I_k} \psi_i$. In its response, the Centre informs rejection price value as for resource μ . On the basis of the value, required resource quantity w_i, and desired effect of its use ψ_i, each automatic machine identifies its own behaviour strategy.

216

To make the problem solution appear as an outcome of automatic machines collective work, it is required to provide for each of them the ability of alternative choice of behavioural strategy. It means that a consumer should benefit not only from the resource use but also from rejecting of it. Making comparison of own benefit (loss), in both cases it is necessary to make decision on strategy selection. Hence, abandoned resource should be paid for.

Assume that payment charged for resource rejection is C_i. It means that a Centre which allocates resource should pay compensation for disconnected consumers in the amount of $\sum_{i=1}^{m}(1-\alpha_i)C_i$. In this context of mutual payments for the resource use and for rejection of it, each automatic machine makes decision in favour of this or that behavioural strategy relying on such a simple rule as

$$\alpha_i' = \begin{cases} 1, & \text{if} \quad \psi_i - \lambda w_i > C_i \\ 0, & \text{if} \quad \psi_i - \lambda w_i < C_i \end{cases} \quad (3)$$

Consider that payment charged for resource rejection and its use efficiency are equal, and decision concerning behavioural strategy selection by automatic machine is random.

After automatic machines have made their choice, they inquire the Centre about resource. The latter controls an execution of limitation (2). Assume that the limitation is executed. If $\left|\sum_{i=1}^{m}\alpha_i' w_i - W\right| \le \inf\{w_i\}$, then the problem is solved by this means: $\alpha_i = \alpha_i'$. If not, resource rejection solution is made by too many automatic machines. Hereupon, given resource limit is underutilized, and the situation is unfavourable for the Centre. So, selected strategies are confirmed only for those who made decision of using the resource ($\alpha_i' = 1$). Those, who rejected the resource ($\alpha_i' = 0$) can not draw confirmation of their choice. The former cut out with their resource portions, and the latter stay to act out remaining resource among themselves in the amount of $W - \sum_{i=1}^{m}\alpha_i' w_i$. If limitation (2) is not executed, then it means that too few automatic machines make decision for rejection of the resource. Hereupon the Centre can not meet the demand as for the resource. That is why selected strategies are confirmed only for those who made

decision for rejection of the resource ($\alpha_i' = 0$). They cut out. The rest of automatic machines stay to act out the resource in the amount of W.

Such a procedure of making one-way decision by a certain part of allocation participants is good as their choice will stay invariable at following stages. Indeed, in case one, the Centre should cut the payment charged for the discarded resource to increase the demand. Hence the inequation $\psi_i - \lambda w_i > C_i$ will be completed during the following stages of the draw (event selection) for those who are still connected to the resource source regardless of other participants behaviour. In case two, to decrease the demand for the resource, the Centre should increase payment charged for those who abandoned the resource. Hence, inequation $\psi_i - \lambda w_i < C_i$ stays invariable during the following stages of the draw (event selection) for those who make decision to be disconnected.

In such a way, making decision by automatic machines as for self-disconnection is multistage procedure. To provide its completion within finite quantity of stages, it is required to apply dividing at each (k^{th}) step of range I_k of participants in k^{th} draw stage (event selection) into two nonempty subsets. If the condition is not met, then dynamic instability may arise within automatic machines collective. The essence is that after one and the same strategy has been selected by all automatic machines, they can not draw its confirmation. They take part again, and all in one select the same strategy once more. The situation cycles over and over again. Dynamic instability hopelessly protracts a procedure of making decision. That conflicts with the interests of the Centre to execute limitation (2) under restricted period. To avoid dynamic instability, the Centre should balance a sum of possible payments charged for rejection of the resource as well as total effect of its use

$$\sum_{i \in I_k} \psi_i - \lambda \sum_{i \in I_k} w_i = \sum_{i \in I_k} C_i \quad (4)$$

Indeed, when (4) is executed, then inequalities of $\forall i \in I_k (\psi_i - \lambda w_i > C_i \lor \psi_i - \lambda w_i < C_i)$ type are excluded. Creation rule for the Centre to charge payment for rejection of the resource follows from (4). In this context the Centre can apply different approaches:

1. "Egalitarianism", when each participant is charged a similar fee regardless of the resource quantity lost by participant

$C_i = C = \sum_{i \in I_k} (\psi_i - \lambda w_i)/N_k$, where N_k is the number of the draw (event selection) participants at k^{th} step.

2. "Linear stimulation" – payment charged for resource disconnection is proportional to its quantity $C_i = \mu w_i$, $\mu = \sum_{i \in I_k} (\psi_i - \lambda w_i)/\sum_{i \in I_k} w_i$.

3. "Nonlinear stimulation" – payment charged for resource disconnection is its quantity nonlinear function.

Also, the Centre may turn to more complex approaches to charge payments for rejection of resource charges; however, studies of adequate model are beyond the scope of the paper.

4 ESTIMATION OF ALGORITHM

Approach of the Centre determines which consumers will use allocated resource. For example, under "egalitarianism", those will be preferred whose benefit from use of resource goes beyond its arithmetic mean value on the number of participants of allocation. However, in this context specific efficiency $\varepsilon_i = \psi_i / w_i$ is not taken into account. If "nonlinear stimulation" approach is applied, then a certain class of consumers may be preferred, for example the ones who need specific preset resource quantity, or some other factors on which charging of payments for rejection of resource depends can be taken into account.

"Linear stimulation" approach prefers those customers who use the resource with maximum specific efficiency

$$\frac{\psi_i}{w_i} - \lambda \geq \sum_{i \in I_k} \psi_i / \sum_{i \in I_k} w_i \qquad (5)$$

Consider design concept of finite automatic machine which implements "linear stimulation" approach. Automation graph is in Figure 2.

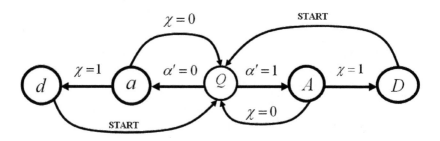

Figure 2. A graph of transition of automation machine taking part in collective allocation of resource

States of automaton and relevant operations are explained in Table 1.

Table 1. States explanation

State		Operation
Term	Explanation	
Q	Initial state	Inquiry forming and sending it to the Centre
a	Preliminary decision on rejection of the resource	Zero
A	Preliminary decision on the resource achieving	Inquiry forming and sending it to the Centre

Continuation of Table 1.

State		Operation
Term	Explanation	
d	Final decision on rejection of the resource	Disconnection, $\alpha = 0$
D	Final decision on the resource achieving	Connection, $\alpha = 1$

Following algorithm sets the automation machine behavior:

1. Message START by the Centre to each automatic machine puts into effect first initiation (transition to a state Q).

2. Under the state, automation machine inquires the Centre (w_i, ψ_i).

3. The Centre backs automation machines the resource value λ, and rejection payment value μ.

4. If $\alpha_i' = 0$ by (3), then automation machines transits to a state (preliminary decision on rejection of the resource).

5. If $\alpha_i' = 1$ by (3), then automation machine transits to A state (preliminary decision on the resource use), and inquires the Centre.

6. After the Centre has had all requests, it calculates and backs automation machines a vector of strategies confirmation

$$[\beta,\gamma] = \begin{cases} 1,1 & if \quad \inf\{w_i\} \geq W - \sum \alpha_i w_i \geq 0, \\ 0,0 & if \quad \inf\{w_i\} < W - \sum \alpha_i w_i \geq 0, \\ 0,1 & if \quad W - \sum \alpha_i w_i < 0 \end{cases}$$

7. Automation machine identifies transition condition χ_i

$$\chi_i = \beta \vee (\alpha_i' \oplus \gamma).$$

If $\chi_i = 1$, then preliminary selected strategy is confirmed, and automation machine transits to its final state (d or D). If $\chi_i = 0$, then preliminary selected strategy can not be confirmed, and automation machine backs to its initial state, and point 2 is implemented.

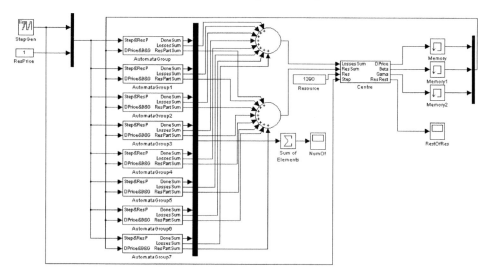

Figure 3. A model of collective behaviour of automatic machines within the problem with "all or nothing" limitations type

The algorithm was analyzed with the help of the model developed in Simulink software (Figure 3).

The model consists of forty automatic machines grouped in five. They differ in resource demands and in effect achieved if it is used

(Figure 4). Each allocation step based on the block of incremented clearing meter is divided into the three stages: resource rejection pricing, preliminary decision making, and formation of strategy confirmation vector, the final decision of automatic machines.

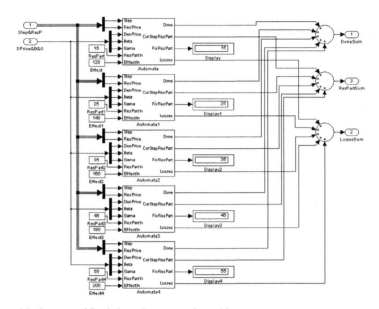

Figure 4. A model of a group of five independent automatic machines

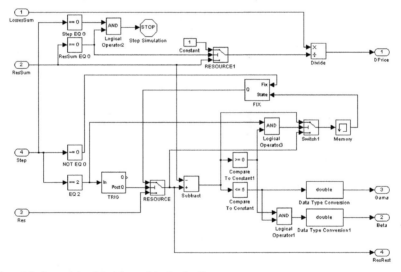

Figure 5. A model of a module of decision making by the Centre

The Centre functions within the model are performed by joint efforts of a scheme which is external for the automatic machines groups. The Centre subsystem block shown in detail in Figure 5 is engaged both in resource rejection pricing and in formation of strategy confirmation vector.

The research purpose is to control the behaviour of proposed decision making algorithm by automatic machines if: $W < \sum_{i=1}^{m} w_i(\Delta t)$,

$$W = \sum_{i=1}^{m} w_i(\Delta t), \quad W > \sum_{i=1}^{m} w_i(\Delta t) \cdot$$

Results of the most representative experiments are deduced in Table 2.

Table 2. The results of the model studies

Experiment	Allocated resource (W)	Desired resource $(\sum_{i=1}^{m} w_i(\Delta t))$	The number of steps/stages to complete allocation	The number of automatic machines which made a decision	The resource rest
$W < \sum_{i=1}^{m} w_i(\Delta t)$	1000	1390	6 / 18	40	36
$W = \sum_{i=1}^{m} w_i(\Delta t)$	1390	1390	5 / 15	40	0
$W > \sum_{i=1}^{m} w_i(\Delta t)$	2000	1390	5 / 15	40	681

Hence, the approach provides making decision on obtaining/rejecting of the resource by all automatic machines within a few steps, and makes the collective possible to adapt for correlation between desired resource quantity and proposed one.

5 ASYMPTOTICAL ESTIMATE OF "LINEAR STIMULATION" APPROACH

To have asymptotical estimation of "linear stimulation" approach from the viewpoint of the problem set (minimum damage caused by resource limitation), consider a situation when such a great number of automatic machines allocate finite quantity of resource W that remained resource tends to zero no matter who of consumers obtain the resource, and who don't.

$$\inf\{w_i\} \geq W - \sum \alpha_i w_i \to 0 \qquad (6)$$

Resource is allocated completely. To characterize efficiency of resource use by consumers, introduce parameter $\varepsilon_i = \psi_i / w_i$ being specific efficiency of the resource use. Rank the consumers in decreasing sequence ε_i and assign a function $\varepsilon(W_j)$ to a set of decreasing values. W_j is a total of resource portions w_i, pretended by automation machines with ε_i values being greater if compared with j^{th} ones. Taking into account extreme, we have

$$\psi(W_j) = \sum_{i=0}^{j} \varepsilon_i w_i \to \int_{0}^{W_i} \varepsilon(W) dW \cdot$$

As the resource is allocated completely, we have

$$\psi(W_0) = \int_{0}^{W_0} \varepsilon(W) dW \qquad (7)$$

where W_0 is allocated quantity of the resource obtained only by consumers having $\varepsilon_i \geq \varepsilon(W_0)$. Consumers ranking according to parameter ε_i is given in Figure 6.

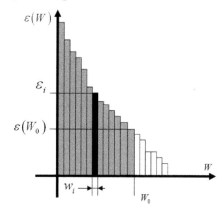

Figure 6. Allocation of the resource among consumers with $\varepsilon_i \geq \varepsilon(W_0)$ parameter (grey-coloured)

Damage due to resource limitation is minimized if $\psi(W_0) = \max \cdot$

Prove that grouping resource-obtaining consumers with $\varepsilon_i \geq \varepsilon(W_0)$ parameter provides damage minimization.

$$F = \int_{W_0}^{W} \varepsilon(W)dW - \lambda(\overline{W} - W_0) =$$

$$\sum_{i=1}^{m} \psi_i - \int_{0}^{W_0} \varepsilon(W)dW - \lambda(\overline{W} - W_0) \to \min$$

Here W_0 is the resource portion pretended by automated machines, $\overline{W} = \sum_{i=1}^{m} w_i$ is maximum resource used by all consumers. It is obvious that $\min\{F\} = \max\{\psi(W_0)\}$ if in the process of allocation, resource value λ stays invariable. To prove that, it is quite sufficient to try to invert a set of allocation participants, i.e. to replace some share of consumers included in a group of those getting resource by those who don't. However, within values of argument $[0, W_0]$, a function $\varepsilon'(W)$ matching the new order will be less than or equal to $\varepsilon(W)$ function. Figure 7 demonstrates such a replacement.

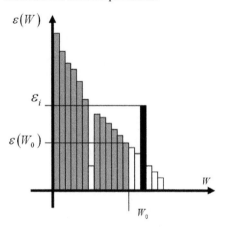

Figure 7. Allocation of the resource among consumers with $\varepsilon_i \geq \varepsilon(W_0)$ parameter (grey-coloured)

As we are interested in asymptotical estimate by (6), then any similar combination will result in a fact that integral obtained due to the replacement will be less than initial one $\psi'(W_0) = \int_{0}^{W_0} \varepsilon'(W) \leq \psi_{\max}$. Hence, selection for a group obtaining resource by $\varepsilon_i \geq \varepsilon(W_0)$ criterion provides asymptotical minimization of damage which was to be proved.

Behaviour of automatic machines determined by (5) meets the demands of the criterion of asymptotical minimum of damage. This implies

that efficiency of above-mentioned allocation algorithm is the greater, the more consumers take part in the resource allocation.

If the number of participant is finite, then allocation being the result of automatic machines self-organization may be nonoptimal as limitation (2) is executed in the form of inequation. Deviation module of practically achievable effect of resource $\psi_{\max}$ use from theoretical maximum complies with the inequation

$$\left| \overline{\psi}_{\max} - \psi_{\max} \right| \leq \varepsilon\left(\sum_{i=1}^{m} \alpha_i w_i \right)\left(W_0 - \sum_{i=1}^{m} \alpha_i w_i \right) \quad (8)$$

As $\psi_{\max} \geq \varepsilon\left(\sum_{i=1}^{m} \alpha_i w_i \right)\sum_{i=1}^{m} \alpha_i w_i$, then dividing left side of inequation (8) into value $\psi_{\max}$, and right side into $\varepsilon\left(\sum_{i=1}^{m} \alpha_i w_i \right)\sum_{i=1}^{m} \alpha_i w_i$, we obtain additional relative divergence value δ of practically achievable effect from real maximum

$$\delta = \frac{\left| \overline{\psi}_{\max} - \psi_{\max} \right|}{\psi_{\max}} \leq \frac{W_0 - \sum_{i=1}^{m} \alpha_i w_i}{\sum_{i=1}^{m} \alpha_i w_i}$$

If the value is greater than certain preset value (e.g. 1%), then the Centre may try to improve a decision by means of changing the approach as for payments charged for resource rejection. For example, it can be done with the help of metagame synthesis (Burkov, 1977).

Constant resource value λ is a distinctive feature of above-mentioned problem. This implies that resource rejection payment falls on its owner as it is in a game with a plus sum. If resource-obtaining consumers pay for rejection of resource (zero-sum game), then the resource value should be variable. Consider resource value creation rule in this case. Total pay rate per rejection of resource is

$$\overline{\sigma} = \mu \sum_{i=1}^{m} (1 - \alpha_i) w_i \quad (9)$$

Options for the sum payment may be provided owing to increased resource value λ'. The increase should execute the equation

$$(\lambda' - \lambda)\sum_{i=1}^{m}\alpha_i w_i = \overline{\sigma} \qquad (10)$$

From the equation with regard to (9) we obtain

$$\lambda' = \frac{\mu\sum_{i=1}^{m}(1-\alpha_i)w_i}{\sum_{i=1}^{m}\alpha_i w_i} + \lambda \qquad (11)$$

If in the process of plus-sum allocation, solution stability is achieved by means of equilibria (4), and asymptotically optimal decision takes no more than m steps, in the case of zero-sum allocation, equilibria (4) can not provide stability due to (11).

6 CONCLUSIONS

A collective of terminal automatic machines implementing the simplest behavioural approaches provides asymptotically optimal solution to the problem of resource allocation with "all or nothing" limitation type.

If solution by automatic machines is that the resource is allocated completely, then the solution is optimum.

There are two quite different approaches in the behaviour of the system of terminal automatic machines: resource allocation with plus sum and zero sum. In case one, automatic machines always reach an agreement within a finite number of steps being no more than the quantity of automatic machines participating in allocation. In case two, the system stability can not be provided. It is required to apply extra efforts to research collective behaviour dynamics and to identify optimum algorithms of collective allocation of the resource.

REFERENCES

Varshavski, V.I. 1973. *Collective behaviour of automatic machines* (in Russian). M.: Nauka: 405.

Burkov, V.N. 1977. *The foundations of mathematical theory of active systems* (in Russian). M.: Nauka: 256.

Petrosian, L.A., Zenkevich, N.A. & Shevkoplias, E.V. 2012. *Games theory* (in Russian). SBR: BHV-Petersburg: 432.

Energy Efficiency Improvement of Geotechnical Systems – Pivnyak, Beshta & Alekseyev (eds)
© 2013 Taylor & Francis Group, London, ISBN 978-1-138-00126-8

An agent-oriented approach to data actualization in databases of electronic atlases

A. Kachanov
State Higher Educational Institution "National Mining University", Dnipropetrovs'k, Ukraine

ABSTRACT: The paper looks at the problems of automatic fetching and preparing the data for electronic atlases. The approach to designing the agent-oriented interfaces for GIS is proposed that is able to solve this task. The structure of the system for automatic data collection and the agent's roles in it are described. The techniques of automated parsing of web pages are discussed.

1. INTRODUCTION

One of the most labor-intensive stages of creating the electronic atlases is the process of collecting, matching, and processing the attributive (statistical) data. It is mainly conditioned by their huge volumes and high diversity. This stage is especially important in the course of creation of a pilot version of the atlas, when fetching the data is performed by experts, who also take part in other development stages. In this paper, the atlas is defined as a system of electronic maps united by a common goal into a complete cartographical product (Busygin et al. 2005). In case of electronic atlases, used in the decision-support process for sustainable management development, their main feature should be the relevance of information (the preferential format of which consists of statistical tables reflecting the ecological, social, and economic features of the regions under study).

Keeping the atlas source data updated should be a central task after creation of the atlas. This process includes collecting and adding new tabular, attributive, and statistical data to the database. In the simplest and least expensive case, the updates are done without creating new thematic sections, just by refreshing and extending time ranges in which the atlas sections are presented.

In practice, however, after the development of electronic cartographical products, the interest of their developers as well as their funding decrease, with rear exceptions of some profitable (Google Earth) or open for free editing social projects (OpenStreetMap). Besides, many electronic atlases are produced in the form of CD-ROMs, in this case, operational updating of informational contents is technically impossible, as in case of paper editions.

Therefore, the issue of automating the process of collecting source data is rather topical. It will make it possible to keep the atlas up to date without considerable efforts.

In Section 2, the objective of the study is formulated and the existing ways of its solution are introduced. Section 3 describes the automation technology of atlas update. Section 4 includes conclusions and directions of further research.

2. FORMULATION OF THE PROBLEM

One of the problems of the process of searching and collecting data for the atlas is diversity of data sources. At present, the fastest and most convenient channel of collecting the data is a global computer network, in particular, the Internet. Specificity of using the Internet is that, from the point of view of atlas creation, it presents itself as a distribution system of heterogeneous data sources. Data search and collection in the Internet can't be performed by using the same algorithm. Each data source requires specific access methods, such as handling and sampling from different databases, processing tabular and PDF documents, obtaining data from web pages, moreover, in each case, the technical details of obtaining data may differ significantly.

Different approaches have been proposed to solve this problem:

- creation of a uniform bank of geospatial thematic data (Krayuknin 2010);
- creation of national and regional infrastructures of spatial data (SDI) (Putrenko 2011; Beshentsev 2010; Karpins'kiy, 2011);
- use of semantic web, including semantic networks and ontology (Grobelny et

al. 2011) and data suitable for computer processing.

Realization of any of these approaches would definitely give a positive result.

It's obvious that after creation of the united infrastructure of spatial data (ISD) and geospatial data banks as well as practical realization of semantic web concepts, it could be possible to simplify, formalize and automate searching and collecting data for the purpose of developing electronic atlases. Unfortunately, at present there is a lack of fully functional systems which collect, accumulate and give access to united databases of geospatial and statistical information at national levels.

In practice, the enumerated approaches, in most cases, are limited to theoretical research and development of recommendations. The main difficulty of their practical implementation lies in huge labor expenses on data processing, moreover, at this stage, a man but not a computer should perform data fetching and processing.

The exception from the rule is the system of Eurostat statistical service of the European Union (www.ec.europa.eu/eurostat) dealing with the collection of statistical information on the EU member-states and the harmonization of statistical methods used by these countries.

This paper proposes a method of solving the problems of collecting and preprocessing the data from heterogeneous sources for its application in the electronic atlas database by using easy to implement in practice the technology of software agents and multi-agent systems.

3. TECHNOLOGY DESCRIPTION

The proposed technology is based on using agents to perform routine tasks of searching and preparing source data and updating and actualization of attributive information in the atlas sections (fig. 1).

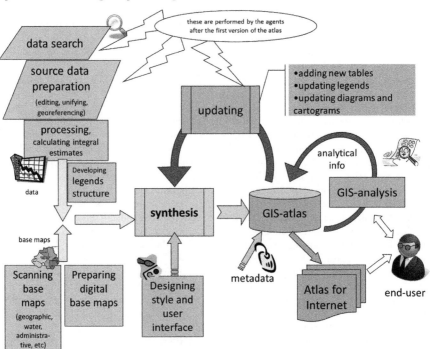

Figure 1. Dataflows in the process of developing and updating electronic atlases

In this case, the agent is a software object (entity) possessing definite artificial intelligence and functioning autonomously to collect data from external data sources. For each data source, separate independent agents, "knowing" technical details of data fetching, are created, such as:

- the language or protocol of database access,
- a method of parsing of a web page and retrieving data from it, filtering HTML-tags,
- recognition of a text and tables in PDF-files,

- use of suitable archiving algorithm or codec,
- etc.

An agent may be not only a software entity, but also a software-hardware entity, for example, for obtaining and preparing data from paper sources (the complex – a scanner and corresponding software for preprocessing and optical character recognition of a text and tables).

Since collecting agents don't perform any analysis or processing of collected data and don't have any knowledge about the whole system structure, their main purpose is to obtain factual data from the source and transfer them to the agent of corresponding specialization for further processing. Separate agents are created for work at the next stages (fig. 2):

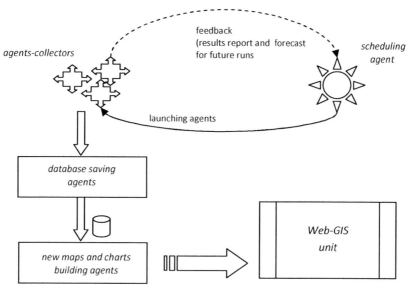

Figure 2. Structure of the agent-based data collection system for electronic atlas

- a data preprocessing agent,
- an agent for database queries and transfers,
- an agent for constructing new maps, charts, and diagrams (as an interface for geoinformational system),
- a scheduler for running the collectors by variable schedule, depending on the activity (of update frequency) of the source.

A scheduling agent doesn't determine the next running time of the agent autonomously. Instead, each agent calculates the next time himself and sends it to the scheduling agent. The algorithm for calculating periodicity makes it possible to optimize the load on data sources and it works in the following way: initially, periodicity default (for example, one week) is set for agents; if after the next run the new data are obtained from the source, the next running time of this collecting agent will be reduced by a few days. If, on the contrary, new data are not found in the source, the survey period will increase. Thus, after several iterations, the agent determines the data

update frequency for the given source himself and chooses the optimal request rate measured from real-time updates to annual updates. Yet, a scheduling agent should take into account the load in the system and can delay the run depending on the priority established for different agents.

Let us consider the development of collecting agents. As it was mentioned above, they can access data storages by using different protocols. When working with databases, the ways of obtaining data are obvious: the use of a corresponding query language or DBMS-specific access schemes, the tabular or hierarchical data ready for further processing are on the output.

In a more general situation, we access the unstructured sources which don't have any special protocols for data fetching. Web pages refer to them. They are currently the main suppliers of data for electronic atlases. The greatest effect from using the agents for data fetching is displayed when working with the data on the Web.

Collecting agents should be configured and trained for each data source separately, thanks to which it is possible to obtain the data in an automatic mode. For automatic analysis and parsing of a web page the following approaches can be distinguished:

- simple parsing using string functions and regular expressions;
- using the Document Object Model (DOM) for HTML document;
- conversion of a HTML document into a "well-formed" XML document and the use of the query language XPath.

XPath parsing is a very convenient and versatile way of analyzing webpages, whereas it's the least time-consuming one. This method, however, has one restriction: a parser can only accept only those XML documents which conform to strict standards of formatting and structure. HTML standards (especially before xHTML 1.0 Strict) were quite loose, so not all HTML documents may be correctly transformed to XML for parsing. That's the reason for using different workarounds and tricks when making XPath queries during the agent training. When a webpage is preprocessed and loaded by XML-parser, we can easily address and query any part of it using particular features like table headers and CSS (Cascade StyleSheets) classed and identifiers, which are widely used to apply formatting to webpage elements. More complex queries can also be composed, such as "getting all table rows after the second section with more than 3 cells in a row and not with specific excluded titles or red marks". This parsing technique, when used in combination with pre-filtering of inconsistent webpages, allows to successfully extract information even from the most unstructured websites, created in a "pre-historic" era without CMS (Content management systems), which usually forces users to adhere to a certain structure and formatting when creating pages.

4. CONCLUSIONS

The proposed approach to designing the agent-oriented interfaces for GIS, based on multi-agent systems, makes it possible to automate the process of searching the attributive data and updating the databases of electronic atlases.

This technology can't make the process of updating the atlas fully automatic since initial training of agents must be performed by an expert programmer. The training consists in the creation of procedures and programs with the help of which the data are fetched from each source. But after training, the data are collected automatically with adaptive tuning of an interval of the next check, and on the basis of the collected data the actualization of corresponding sections of the atlas is performed. This makes the self-sufficient auto-updating subsystem of data storage easily implemented in the development and update of electronic atlases.

In the perspective, the agent-oriented technology may be further improved by using additional automation of the process of searching for new data sources, in particular, by analyzing existing hyper-texts and highlighting specific keywords which can be used for finding websites, "similar to" already known.

REFERENCES

Busygin B.S., Kachanov A.V., Sarycheva L.V. 2005. *Creation of electronic atlas of Ukraine's regions sustainable development* (in Russian). Scientific works of Vernadsky Taurida National University, Vol. 18(57), Issue 1: 9-15.

Krayukhin A.N. 2010. *The role and place of thematic cartography in the System of State information resources* (in Russian). Irkutsk: Sochavy Geography Institute: Proceedings of the IX scientific conference about thematic cartography, Vol. 1: 5-7.

Putrenko V.V. 2011. *World experience in organizing thematic information in geospatial data infrastructures* (in Ukrainian). Kyiv: Development of thematic infrastructure of geospatial data in Ukraine: Scientific works: 133-138.

Beshentsev A.N. 2010. *Mapping geospatial data infrastructures* (in Russian). Irkutsk: Sochavy Geography Institute: Proceedings of the IX scientific conference about thematic cartography, Vol. 1: 25-28.

Karpins'kiy Yu.O. 2011. *From infrastructure of cartography production to geospatial data infrastructures* (in Ukrainian). Kyiv: Development of thematic infrastructure of geospatial data in Ukraine: Scientific works: 39-61.

Grobelny P. & Pieczyński A. 2011. *Results of research on method for intelligent composing thematic maps in the field of Web GIS.* Lecture Notes in Artificial Intelligence: Computational Collective Intelligence, Technologies and Applications, LNAI 6922, Springer, Berlin, Heidelberg: 264—274.

Investigating the thermal condition of slagheap landscapes according to the data of Landsat-TM/ETM+

K. Sergieieva

State Higher Educational Institution "National Mining University", Dnipropetrovs'k, Ukraine

ABSTRACT: The distribution of brightness temperature fields of slagheap landscapes is investigated for the city of Donetsk (Ukraine) and adjoining territories, according to the information of multi-temporal satellite imagery Landsat-TM/ETM+ in the infrared (thermal) range of the electromagnetic spectrum. The methodology is proposed for detecting temperature anomalies of smoldering and burning slagheaps as well as for quantifying their impact on the environment.

1 INTRODUCTION

Slagheap landscapes are industrial anthropogenic landscapes, the main elements of which are tapered or conical waste rocks, formed as a result of underground mining, in particular, of coal mining. Coal slagheaps contain 5-15% of coal residues inclined to spontaneous combustion under certain conditions (high moisture and access of air oxygen).

Today, in Ukraine's part of the Donetsk Coal Basin there are more than one thousand slagheaps and a third of them are burning ones. The surface of smoldering and burning slagheaps pollutes the atmosphere of adjoining territories by tons of dust and dangerous substances, including SO_2, NO, CO, CH_4, N_2O, CO_2 etc. (Smirnyy et al. 2006). According to the current state sanitary rules and norms of Ukraine, each slagheap must be surrounded by a sanitary protection zone of at least 300 m from residential areas. However, these requirements are frequently not observed as a result of which products of burning slagheaps have an extremely negative impact on people's health. Detection of slagheaps, that provide intense influence on the natural environment, is a priority task in the system of measures directed on the liquidation of heaths of fire (Busygin et al. 2010).

Detecting the sites of uncontrollable processes of burning and decaying the coal-bearing rocks of slagheaps is a priority challenge for the world coal basins, the largest of which are:

− coal deposits of Central and Northern China (Chatterjee 2006; Gangopadhyay et al. 2006; Kuenzer et al. 2007);

− Wuda coalfield, Inner Mongolia (Zhang et al. 2007);

− coal deposits of the United States of America (Stracher et al. 2004), etc.

It is known that ground-based measurements are not effective for coal fire detection and monitoring due to their relatively low efficiency and high cost of works. Therefore, to solve the problems of detection, mapping and monitoring of the sites of burning coal-bearing rocks (Chatterjee 2006; Gangopadhyay et al. 2006), researchers from the whole world use the Earth remote sensing data from space (ERS data). The ERS data are also used for fire monitoring of subsurface coal seams (Latifovic et al. 2005; Csiszar et al. 2008; Ranjan et al. 2005), detecting temperature anomalies of coal fires (Zhukov et al. 2006; Gautam et al. 2008), mapping of active surface fires and burned areas (Ressl et al. 2009). The advantages of such data are high spatial coverage and possibilities of multiple or repeated imagery of the territory.

Smoldering and burning slagheaps are characterized by increased values of temperature over the heaths of fire. The ERS data in the thermal range of the electromagnetic spectrum (8–15 μm) make it possible to estimate the brightness temperature and detect processes characterized by intensive heat release, including the process of decay and burning coal-bearing rocks of slagheaps. In particular, the data of the Landsat-TM band 6 represent the greatest interest among the free-access thermal imageries (Landsat-ETM+ bands 6.1, 6.2) in the spectral range 10.4-12.5 μm. Their principal advantage, in comparison with other free-available materials, is higher spatial resolution – respectively, 120 m and 60 m. The main disadvantage of the Landsat-TM/ETM+ data is a low temporal resolution of clear-sky images. The latter circumstance doesn't make it possible to use traditional methods of time series forecasting and mathematical modeling to investigate the thermal state of slagheap landscapes. Such methods suggest observance of the principle of completeness and equality of intervals between imagery dates, as well as presence of a deterministic component in the time series. Violation of these assumptions requires search of new approaches to the investigation of slagheaps and detection of heaths of fire using multi-temporal samples of space images.

Thus, this paper is aimed at the establishment of a methodology for processing multi-temporal space imagery in the thermal range of the electromagnetic spectrum in order to improve effectiveness of decision support in the tasks of investigating the thermal condition of slagheap landscapes and detecting burning slagheaps.

2 METHODOLOGY

The methodology for investigating the thermal condition of slagheap landscapes has been established. It includes: the stages of preliminary impact assessment of slagheaps on the environment of the adjoining areas within buffer zones, analysis of statistical characteristics of multi-temporal temperature images, detection of temperature anomalies on the basis of multiple seasonal data, categorization of slagheaps as burning and non-burning based on the information about temperature anomalies, as well as calculation of quantitative characteristics for temperature anomalies.

2.1 Input data and test site

In this study, the site of the Donetsk Coal Basin (Figure 1a) including the city of Donetsk and adjoining areas (sizes – 40.5 × 26 km, area – 1053 km^2) is investigated according to the data of the thermal bands of 15 Landsat-TM/ETM+ images. Acquisition dates: 03.05.2000, 26.10.2000, 04.04.2001, 09.05.2002, 29.08.2002, 28.05.2003 (Landsat-ETM+) and 02.05.1985, 03.04.1986, 15.05.2007, 21.06 .2009, 07.07.2009, 09.09.2009, 08.06.2010, 12.09.2010, 27.06.2012 (Landsat-TM).

To assess the reliability of temperature mapping results the information on the condition of 96 Donetsk slagheaps (control sample) has been used (Figure 1b). It is presented on the website of the Department of Environmental Safety of the Donetsk City Council (http://www.donmaps.org.ua).

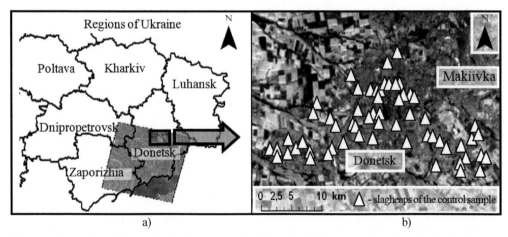

a)   b)

Figure 1. a) Location of the test site; b) location of control sample slagheaps on the Landsat-TM band 6 image registered on June 27, 2012.

2.2 Assessment of the impact of slagheaps on the environment of the adjacent territories

For the site being investigated as a result of classification of the multispectral image Landsat-TM (acquisition date: 27.06.2012), four types of land cover were identified: buildings and facilities, open soil, water objects and vegetation.

Around contours of slagheaps, outlined according to Google data, buffer zones with the radius of 300 m were built and density of different types of land cover within buffer zones of slagheaps was analysed (Table 1).

It was found that 11.54% of the area of slagheap buffer zones of the test site are occupied by industrial and residential buildings and facilities which are exposed to the negative environmental impact. This fact confirms the expediency of detecting burning slagheaps of the investigated site which pose special danger to public health. The presence of vegetation within buffer zones of slagheaps (58.26% of the area of buffer zones) reduces a negative impact of slagheaps on the environment. However, presence of vegetation in some cases complicates the task of detecting temperature anomalies of burning slagheaps.

Table 1. Statistics of sites with different land cover types

Statistics \ Land cover types	Buildings and facilities	Open soil	Water objects	Vegetation
The area occupied by objects of the class, km^2	121.94	246.51	10.00	674.55
The part of the area of a test site, %	11.58	23.41	0.95	64.06
The area occupied by objects of the class within buffer zones of slagheaps, %	14.07	72.82	0.06	392.99
The share of objects located within buffer zones of slagheaps, relative to the total number %	11.54	29.54	0.66	58.26

2.3 Analysis of statistical characteristics of temperature images

One of the initial stages of implementing the methodology for studying slagheap landscapes was analysis of distribution types and basic statistical characteristics of multi-temporal values of brightness temperature for Landsat-TM/ETM+ images. The results of the analysis made it possible to select the rule for detecting the sites with increased temperatures – temperature anomalies of coal fires. For this purpose, the "raw" data of thermal satellite imagery were converted into the sensor incoming radiation and then recalculated into the values of brightness temperature (°C) by the Plank formula (Gangopadhyay et al. 2006):

$$L_\lambda = \frac{L_{max\,\lambda} - L_{min\,\lambda}}{Qcal_{max} - Qcal_{min}}(Qcal - Qcal_{min}) + L_{min\,\lambda} \quad (1)$$

$$T = \frac{K_2}{\ln\left(\left(K_1/L_\lambda\right)+1\right)} - 273,15 \quad (2)$$

where:

$L_\lambda, L_{min\,\lambda}, L_{max\,\lambda}$ – amount of radiation reaching the sensor; $Qcal, Qcal_{min}, Qcal_{max}$ – calibrated pixel value; T – brightness temperature; K_1, K_2 – calibration constants.

Figure 2 shows an image and a histogram of brightness temperature values for the Landsat-TM thermal image registered on June 27, 2012. It was found that all Landsat-TM/ETM + thermal images are characterized by a close to normal distribution of temperature values. Temperature histograms for all images are symmetric and unimodal. Mean, mode and median values of brightness temperature are similar, and the values of coefficient of variation (CV) do not exceed 0.3. This leads to the conclusion about a normal type of distribution temperature data and detection of anomalies by the "three-sigma rule", according to which about 68% of all possible temperature values fall into the range *Mean ± 1Std. Dev.*, about 95% – into the range *Mean ± 2Std. Dev.* and almost all temperature values are enclosed in the range *Mean ± 3Std. Dev.*, where *Std. Dev.* is a temperature standard deviation. In this study, the anomalous brightness temperature values were outside the range (*Mean ± 1Std. Dev.*) for different moments of imageries.

Values of pair correlation coefficients for all possible combinations of thermal images vary from 0.14 to 0.81. Such instability of linear relations between multi-temporal temperature data can be explained by seasonal influences, different weather conditions and date of registration of images, and as a result by differences in vegetation cover and spatial distribution of temperature within the test site.

231

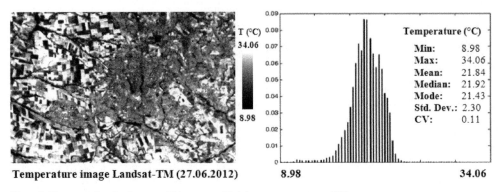

Temperature image Landsat-TM (27.06.2012)

Temperature (°C)	
Min:	8.98
Max:	34.06
Mean:	21.84
Median:	21.92
Mode:	21.43
Std. Dev.:	2.30
CV:	0.11

Figure 2. The example of an image and histogram of brightness temperatures (°C).

2.4 Detection of temperature anomalies

Combustion processes of coal-bearing rocks of slagheaps are not continuous over time and can occur in different parts of slagheaps, depending on the external conditions (presence of fresh rock, precipitation, access of oxygen and air etc.). The possibility of satellite detection of temperature anomalies, connected with the processes of burning coal remains of slagheaps, is also influenced by the image registration date, which determines availability and density of the vegetation cover that reduces a thermal impact of slagheaps on the environment.

Due to the lack of explicit regularities between multi-temporal images of surface temperature, the initial multitude of thermal images Landsat-TM/ETM+ was divided into 4 samples to solve the task of detecting temperature anomalies. Among them 3 are seasonal, and one includes all 15 images:

– sample 1: all 15 images;
– sample 2: 9 images registered in May and June;
– sample 3: 3 images registered in August and September;
– sample 4: 3 images registered in April and October.

The site being investigated is represented by the images of sample 3. It is characterized by the presence of thicker vegetation in comparison with sample 2 and is more affected by intensive seasonal solar radiation. On the images of sample 4, the test site is characterized by sparse or nearly absent vegetation.

For each sample, negative ($T < Mean – 1Std.$ $Dev.$) and positive ($T > Mean + 1Std. Dev.$) temperature anomalies were detected. Negative anomalies correspond to the location of shaded areas, parks and water objects. Location of positive temperature anomalies is typical for areas heated by solar radiation – open soils, synthetic surfaces (building roofs, industrial sites) and burning slagheaps. Positive temperature anomalies are mostly noticeable for samples 3 and 4 – for images registered in April, August, September and October. Heating of soil in samples 3 and 4 is caused by intense solar radiation, and in sample 4 – also by the lack of vegetation. In particular, positive temperature anomalies were detected on the territories of the Donetsk Metallurgical Plant, the Donetsk Cotton Plant, the "Tochmash" company and other industrial areas, fields and slagheaps. Slagheaps were classified as burning and non-burning on the basis of information on the presence or absence of temperature anomalies within slagheap images. Classification accuracy was assessed by comparing the information on the presence of temperature anomalies on slagheap images with the information on the state of combustion in the control sample and calculation of a general classification error as well as errors of the first and second type. The largest classification error was 26.2% for sample 1 because of the widest range of acquisition dates. Samples 2 and 4 are characterized by the highest percentage of correctly classified slagheaps – 79.8% and 81.0% respectively (Table 2).

Table 2. Classification results of slagheaps

Classification results	Sample 1	Sample 2	Sample 3	Sample 4
Classification error (%)	26.2	20.2	22.6	19.1
Type I error (%)	1.2	4.8	10.7	7.1
Type II error (%)	25.0	15.5	11.9	11.9
Correctly classified slagheaps (%)	73.8	79.8	77.4	81.0

Mismatches of classification results of slagheaps, according to the data about positive temperature anomalies of separate image samples with priori data of ground-based observations is often explained by the presence of vegetation cover. As to the burning slagheaps of the mines "Petrovsky", 11-bis, "Trudovskaya", "Abakumov", 17-bis, "Kuibyshev", "Panfilov", "Octyabrskaya", "Proletar", "Gorkogo", "Centralno-zavodskaya", "Putilovskaya", "Zasyadko" and "Krasnaya Zvezda" no temperature anomalies were detected on different image samples. Besides, in some cases not covered or weakly covered with vegetation slagheaps were classified as burning (slagheaps of the mines № 2-16, 17-bis, № 16-17 "Evdokievka", "Sorokinskaya", № 8 "Chulkova", "Sotsdonbass newspaper", № 12-18 "Pravda" and № 6 "Capitalnaya").

According to the images of sample 1, slagheaps of the mines № 10, "Trudovskaya", № 5, № 6, № 19, № 30, № 31 "Gorkogo", № 3, "Shvernika", № 4-17, № 4 "Livenka", № 1-2 "Smolyanka", "Octyabrskaya", "Vetka Naklonnaya", № 4 "Gorkogo", "Vladimir", № 12 "F. Cohna", "Maria", "Leningradka", "Putilovskaya", "Centralno-zavodskaya", № 16 "Mushketovskaya", "Mushketovskaya-Naklonnaya" and "Chumakovskaya" were correctly classified as non-burning. Slagheaps of mines № 29, "Yuznaya", № 9 "Capitalnaya", "Zaperevalnaya", "60th anniversary of the USSR", № 12 "Naklonnaya", "Chelyuskintseu", "Kalinina", "Chumakovkaya" were correctly classified as burning, and 3 of them are active (Figure 3).

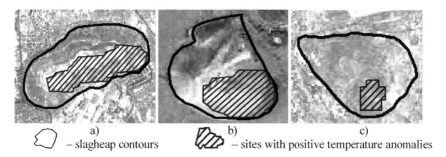

a)   b)   c)

⬭ – slagheap contours   ⬬ – sites with positive temperature anomalies

Figure 3. Images of high spatial resolution from Google Internet-service: slagheaps of the mines: (a) "Zaperevalnaya", (b) № 12 "Naklonnaya" and (c) № 29 with marked location of temperature anomalies.

2.5 Calculation of quantitative characteristics of temperature anomalies

Calculation of quantitative characteristics of temperature anomalies of burning slagheaps makes it possible to assess the degree of expressiveness or intensity of combustion processes and differentiate slagheaps by their impact on the surrounding environment of the adjacent territories. In the study, three traditional characteristics were used (Schwarz et al. 2011):

P_1: the mean temperature within a buffer zone of a slagheap;

P_2: the mean temperature within a test site;

P_3: the mean temperature of water objects.

It was established that the highest values of P_1-P_3 characteristics are inherent to temperature anomalies of "Zaperevalnaya" mine and №12 "Naklonnaya" mine slagheaps. According to P_2-P_3 characteristics the least impacts on the environment slagheap of mine № 29 (Figure 3).

CONCLUSIONS

Investigation of the temperature condition of Donetsk slagheap landscapes and of adjacent territories by means of the multi-temporal thermal imagery Landsat-TM/ETM+, registered in 1985-2012, revealed spatial and temporal temperature anomalies of the burning slagheaps with the accuracy of up to 81.0% and made it possible to carry out quantitative assessment of the impact intensity of burning sites on the environment of the adjoining territories by calculating the ratio of the mean temperature of a slagheap anomalous site to the mean temperature of surrounding sites with different land cover types (characteristics P_1-P_3). It was established that the possibility of detecting positive thermal anomalies is conditioned by the presence or absence of the vegetation cover and the range of acquisition dates for image samples.

The temperature mapping results are designed for zoning of urban territories by the level of environmental safety and for improving the effectiveness of decision support in the tasks of monitoring and elimination of the consequences of the processes of burning slagheaps.

REFERENCES

Smirnyy M.F., Zubova L.G., Zubov A.P. 2006. *Environmental security of the Donbass slagheap landscapes: Monograph* (in Russian). Moscow: Publishing House of the VNU V. Dalya: 232.

Busygin B., Garkusha I., Sergieieva K. 2010. *Using the space survey data for fire objects monitoring of the Donetsk coal basin.* International Symposium on Environmental Issues and Waste Management in Energy and Mineral Production SWEMP 2010, 12th: Proceedings. Prague: Lesnicka prace: 25-30.

Chatterjee R.S. 2006. *Coal fire mapping from satellite thermal IR data – A case example in Jharia Coalfield, Jharkhand, India.* ISPRS Journal of Photogrammetry & Remote Sensing, Issue 60: 113–128.

Gangopadhyay P.K., Lahiri-Dutt K., Saha K. 2006. *Application of remote sensing to identify coalfires in the Raniganj Coalbelt, India.* International Journal of Applied Earth Observation and Geoinformation, Issue 8: 188–195.

Kuenzer C., Zhang J., Tetzlaff A., Dijk P., Voigt S., Mehl H., Wagner W. 2007. *Uncontrolled coal fires and their environmental impacts: Investigating two arid mining regions in north-central China.* Applied Geography, Issue 27: 42–62.

Zhang J., Kuenzer C. 2007. *Thermal surface characteristics of coal fires 1 results of in-situ measurements.* Journal of Applied Geophysics, Issue 63: 117–134.

Stracher G.B., Taylor T.P. 2004. *Coal fires burning out of control around the world: thermodynamic recipe for environmental catastrophe.* International Journal of Coal Geology, Issue 59: 7– 17.

Latifovic R., Fytas K., Chen J., Paraszczak J. 2005. *Assessing land cover change resulting from large surface mining development.* International Journal of Applied Earth Observation and Geoinformation, Issue 7: 29–48.

Csiszar I.A., Schroeder W. 2008. *Short-Term Observations of the Temporal Development of Active Fires From Consecutive Same-Day ETM+ and ASTER Imagery in the Amazon: Implications for Active Fire Product Validation.* IEEE Journal of Selected Topics in Applied Earth Observations and Remote Sensing, Vol. 1, Issue 4: 248–253.

Ranjan M.T., Bhattacharya A., Kumar V.K. 2005. *Coal-fire detection and monitoring in Raniganj coalfield, India – A remote sensing approach.* Current Science, Vol. 88, Issue 1: 21–24.

Zhukov B., Lorenz E., Oertel D., Wooster M., Roberts G. 2006. *Spaceborne detection and characterization of fires during the bi-spectral infrared detection (BIRD) experimental small satellite mission (2001–2004).* Remote Sensing of Environment, Issue 100: 29 – 51.

Gautam R.S., Singh D., Mittal A., Sajin P. 2008. *Application of SVM on satellite images to detect hotspots in Jharia coal field region of India.* Advances in Space Research, Issue 41: 1784–1792.

Ressl R., Lopez G., Cruz I., Colditz R.R., Schmidt M., Ressl S., Jiménez R. 2009. *Operational active fire mapping and burnt area identification applicable to Mexican Nature Protection Areas using MODIS and NOAA-AVHRR direct readout data.* Remote Sensing of Environment, Issue 113: 1113–1126.

Schwarz N., Lautenbach S., Seppelt R. 2011. *Exploring indicators for quantifying surface urban heat islands of European cities with MODIS land surface temperatures.* Remote Sensing of Environment, Issue 115: 3175–3186.